Springer Proceedings in Complexity

Springer Proceedings in Complexity publishes proceedings from scholarly meetings on all topics relating to the interdisciplinary studies of complex systems science. Springer welcomes book ideas from authors. The series is indexed in Scopus.

Proposals must include the following:

- name, place and date of the scientific meeting
- a link to the committees (local organization, international advisors etc.)
- scientific description of the meeting
- list of invited/plenary speakers
- an estimate of the planned proceedings book parameters (number of pages/articles, requested number of bulk copies, submission deadline)

Submit your proposals to: Hisako.Niko@springer.com

Andreia Sofia Teixeira · Federico Botta ·
José Fernando Mendes · Ronaldo Menezes ·
Giuseppe Mangioni
Editors

Complex Networks XIV

Proceedings of the 14th Conference on Complex Networks, CompleNet 2023

Springer

Editors
Andreia Sofia Teixeira
Faculty of Sciences
University of Lisbon
Lisbon, Portugal

José Fernando Mendes
Department of Physics
University of Aveiro
Aveiro, Portugal

Giuseppe Mangioni
Department of Electrical, Electronic
and Computer Engineering
University of Catania
Catania, Italy

Federico Botta
Department of Computer Science
University of Exeter
Exeter, UK

Ronaldo Menezes
Department of Computer Science
University of Exeter
Exeter, UK

ISSN 2213-8684 ISSN 2213-8692 (electronic)
Springer Proceedings in Complexity
ISBN 978-3-031-28275-1 ISBN 978-3-031-28276-8 (eBook)
https://doi.org/10.1007/978-3-031-28276-8

This Springer imprint is published by the registered company Springer Nature Switzerland AG
The registered company address is: Gewerbestrasse 11, 6330 Cham, Switzerland

Contents

Brain's Dynamic Functional Organization with Simultaneous EEG-fMRI Networks

Francisca Ayres-Ribeiro[1,2,3](✉), Jonathan Wirsich[4], Rodolfo Abreu[5,6], João Jorge[7,8], Andreia Sofia Teixeira[3], Alexandre P. Francisco[2], and Patrícia Figueiredo[1]

[1] ISR-Lisboa and Department of Bioengineering, Instituto Superior Técnico, Universidade de Lisboa, Lisboa, Portugal
francisca.ayres.ribeiro@tecnico.ulisboa.pt
[2] INESC-ID and Instituto Superior Técnico, Universidade de Lisboa, Lisboa, Portugal
[3] LASIGE, Faculdade de Ciências, Universidade de Lisboa, Lisboa, Portugal
[4] EEG and Epilepsy Unit, Department of Clinical Neurosciences, University of Geneva, Geneva, Switzerland
[5] Coimbra Institute for Biomedical Imaging and Translational Research (CIBIT), Universidade de Coimbra, Coimbra, Portugal
[6] Institute for Nuclear Sciences Applied to Health (ICNAS), Universidade de Coimbra, Coimbra, Portugal
[7] Laboratory for Functional and Metabolic Imaging, École Polytechnique Fédérale de Lausanne (EPFL), Lausanne, Switzerland
[8] Swiss Center for Electronics and Microtechnology (CSEM), Systems Division, Neuchâtel, Switzerland

Abstract. The brain's functional networks can be assessed using imaging techniques like functional magnetic resonance imaging (fMRI) and electroencephalography (EEG). Recent studies have suggested a link between the dynamic functional connectivity (dFC) captured by these two modalities, but the exact relationship between their spatiotemporal organization is still unclear. Since these networks are spatially embedded, a question arises whether the topological features captured can be explained exclusively by the spatial constraints. We investigated the global structure of resting-state EEG and fMRI data, including a spatially informed null model and found that fMRI networks are more modular over time, in comparison to EEG, which captured a less clustered topology. This resulted in overall low similarity values. However, when investigating the community structure beyond spatial constraints, this similarity decreased. We show that even though EEG and fMRI functional connectomes are slightly linked, the two modalities essentially capture different information over time, with most but not all topology being explained by the underlying spatial embedding.

Keywords: EEG-fMRI · Connectomics · dFC · Community analysis

1 Introduction

Brain activity is believed to be organized into functional networks, reflecting the dynamic coupling between brain regions and the continuous exchange of infor-

A. S. Teixeira et al. (Eds.): CompleNet 2023, SPCOM, pp. 1–13, 2023.
https://doi.org/10.1007/978-3-031-28276-8_1

mation throughout the whole brain [19]. This coupling is known as functional connectivity. Characterizing the dynamic behaviour of these networks and their topology might be key to increase the understanding of the brain's complex activity, its spatiotemporal organization and, possibly, provide biomarkers for neurological and psychiatric diseases [9,29].

These functional networks can be defined using different imaging techniques like functional Magnetic Resonance Imaging (fMRI) and electroencephalography (EEG), that allow the characterization of time-varying activity in the whole brain. However, these techniques have distinct temporal and spatial resolution and are sensitive to different physiological changes associated with neuronal activity [21]. FMRI measures brain activity indirectly and is based on changes in the blood flow, consisting in a blood-oxygen-level-dependent (BOLD) signal. These changes, however, come slow and with a significant delay [28]. In contrast, EEG allows the direct measurement of transient brain electrical dipoles generated by neuronal activity [21] - with the use of scalp electrodes -, having high temporal precision. Even though both reveal the brain's dynamic behaviour, it is still not entirely known how the two are correlated, i.e., what is the relationship between the hemodynamic response and electric neuronal activity, and whether they capture the same information or not [24].

In recent years, several studies have analysed functional connectivity by combining simultaneously acquired EEG-fMRI recordings, in resting-state, with the objective of establishing a correlation or link between the two and also to take advantage of their complementarity [1,8,26,33,38]. Moreover, this type of analysis can provide richer characterization of the spatiotemporal organization of the brain activity. However, it is still missing a comparative analysis between these two modalities functional networks by investigating their topology over time. Such analysis can be done considering a graph theory framework that allows the brain functional systems to be modeled as complex networks [9]. In this context, the functional networks and their dynamic topology can be studied by analysing their global properties, such as their community structure [5,18], which looks at the organization of the network into modules reflecting coherent activity between different brain regions.

Furthermore, since these networks are spatially embedded, the question arises whether the topological features captured can be explained by impositions determined by the brain's underlying structure [31], in order to minimize energy costs of maintaining such connections [32], or if there is still some functional synchronization deviating from these proximity constraints. Some studies have explored this spatial effect in structural networks [31,32] and, more recently, in the community structure of functional networks, distinguishing the influence of short- and long-distance connections [16].

Therefore, this study intends to fill the gap in the present literature by performing a comparative network analysis with EEG and fMRI dynamic functional connectivity (dFC) data, on a global level, by means of a community analysis. For that, several established approaches were used, such as the Louvain algorithm [6] for the extraction of modules of coordinated activity, as well as its multiplex version [25], here applied to find partitions combining EEG and fMRI for the first time. Moreover, with hopes of exploring the influence of space in the topology, it was investigated the functional networks topology beyond these spatial constraints. Hence, a new approach was applied - a modified version of the Louvain algorithm [11], that includes a degree constrained spatial null model in the modularity definition.

2 Materials and Methods

2.1 Data Acquisition and Preprocessing

The dataset used in this work consists in simultaneous EEG-fMRI recordings acquired during rest in the scope of a previous project [20], using a 7T MRI scanner along with a 64-channel EEG system, involving 9 healthy subjects (4F, 22–26 yrs). The data preprocessing and brain segmentation was done according to [35]. Moreover, the BOLD timeseries were bandpass-filtered at 0.009–0.08 Hz, while the EEG signals were filtering at 0.3–70 Hz and segmented as a multiple of the Repetition Time (TR) of the fMRI acquisition (TR 1s).

2.2 Construction of Functional Networks

In order to analyse and compare the EEG and fMRI functional networks' topology, a graph representation was used. The nodes were set as 68 regions of interest defined by the Desikan(-Killiany) atlas [14], for both modalities. This parcellation was chosen considering the number of EEG channels and the optimal parcellation size to capture independent EEG signals according to [17]. Moreover, to guarantee this spatial alignment between modalities, the EEG data was subjected to a source reconstruction procedure, using the Tikhonov-regularized minimum norm [2], as described in [35].

The edges, on the other hand, were defined by functional connectivity matrices obtained for each time point (TR) using phase coherence for fMRI and imaginary part of coherency for EEG [27]. The first constitutes an instantaneous connectivity measure, based on phase synchronization, which was estimated using an adaptation of Cabral et al.'s implementation[1] [10], while the second was obtained using the Brainstorm function *bst_ cohn.m* (according to the Brainstorm 2018 implementation, 'icohere' measure[2]) as described in [35]. Furthermore, the imaginary part of coherency estimation was averaged for the 5 canonical frequency bands: delta δ (1–4 Hz), theta θ (4–8 Hz), alpha α (8–12 Hz), beta β (12–30 Hz), gamma γ (30–60 Hz).

To guarantee temporal equivalence between EEG and fMRI functional networks, a motion scrubbing step was taken, excluding, for both modalities, the time points where excessive motion was detected on the EEG data. In addition, as the BOLD time-series have an intrinsic delay with respect to the EEG data

[1] https://github.com/juanitacabral/LEiDA.
[2] http://neuroimage.usc.edu/brainstorm.

due to the different nature of the two signals, this temporal shift was estimated and taken into account in aligning the two modalities' networks. This was done using a resting-state hemodynamic response function (HRF) deconvolution toolbox[3] [37], leading to 3–4seconds delay, depending on the subject.

Finally, to remove any spurious connectivity arising from noise or artifacts typical on this type of dataset, the functional networks were thresholded. The choice of an adequate proportional threshold was made using a data-driven percolation approach [7], i.e., by finding the percentage of edges necessary for each time point to avoid the collapse of the giant component, which guarantees the network's structure integrity. Expecting fluctuations in activity over time, resulting in more or less structured networks [5], it was selected the median value of said percentage of edges to be kept, resulting in the following threshold values: 11%, 7.5%, 6.6%, 7.0%, 6.1%, 7.0% and 6.5% for fMRI and EEG delta, theta, alpha, beta and gamma frequency bands, respectively.

2.3 Community Analysis

Community analysis was performed to characterize both EEG and fMRI functional networks on a macro-scale, to explore their potential similarity over time and also to investigate the possible influence of the proximity constraints in the modular structure captured.

Global Structure Statistical Significance. First, the global topology of these functional networks was analysed over time, in comparison to a rewiring null model, using three metrics: clustering coefficient, average path length and modularity (computed using NetworkX's functions). This allowed for the selection of time instances associated to functional networks whose global structure was statistically significant. In concrete, this statistical testing step was done for all metrics by generating 100 surrogates for each corresponding network and selecting the time instances with $p < 0.05$. Subsequently, the selection of time points was intersected with respect to the three metrics and for both modalities, considering each frequency band independently. This led to a unique set of time instances corresponding to statistically significant structured networks, for EEG and fMRI simultaneously. Moreover, since the topology observed may be due to spatial constraints, this analysis was performed again with respect to a degree constrained spatial null model [11], resulting in a second set of time points deviating from what was expected by the influence of space.

Louvain Algorithm. With the goal of identifying modules of synchronized activity, potentially similar between modalities, the community structure of these functional networks was analysed. This was done using the Louvain algorithm for the set of time points previously selected with respect to the rewiring null model. To explore the potential correlation between the two modalities regarding to the modular structure captured, the communities extracted were compared

[3] https://github.com/compneuro-da/rsHRF.

for all selected time points, using the Normalized Mutual Information (NMI) metric, with values between 0 and 1.

Modified Louvain Algorithm. Considering the influence of the spatial constraints in the functional networks topology, it was desirable to check if some modular configuration present emerged from functional necessity and not just as a consequence of space proximity. With this objective, the community structure beyond these spatial constraints was investigated. This was done using the modified Louvain algorithm[4] [11], for all selected time points obtained previously with respect to the degree constrained spatial null model. Additionally, the communities extracted with and without the spatial influence were compared over time using the NMI metric. Finally, the EEG and fMRI communities extracted while regressing out the influence of space were compared for all time points, again using the NMI metric. This was done to distinguish and quantify the influence of the spatial constraints in the alignment of the two modalities and also to verify if there was still some similarity beyond that, reflecting the underlying synchronous activity captured by the two.

Multiplex Louvain Algorithm. To investigate if combining EEG and fMRI information would lead to new and improved results, a multilayer version of the Louvain algorithm [25] was used to extract communities common to both modalities (using i-graph's louvain package *find_partition* function), for all time points selected using the rewiring null model. This algorithm was applied as a multiplex case, i.e., where all the layers share the same node set, since there is a spatial equivalence between EEG and fMRI networks. With this, the improved modularity was estimated for the combined multiplex EEG-fMRI network, for each frequency band. In parallel, it was computed the modularity associated to these communities when isolating the two layers, to check if this optimization procedure led to different partitions than the ones obtained as described in Sect. 2.3). Additionally, these values were compared with the equivalent ones for the degree constrained spatial null model of both modalities.

3 Results and Discussion

3.1 Global Structure Statistical Significance

The global structure of EEG and fMRI functional networks was analysed, comparing global metrics values with a rewiring and a spatial null model. Fig. 1 illustrates the temporal variation of the modularity values (chosen between the three metrics computed as they showed similar behaviour) in comparison to each null model. Table 1 summarizes the percentage of time points deviating from these null models, averaging for all subjects.

From these results, it is noticeable an oscillation over time, for both modalities, which is not surprising considering that brain functional connectivity tends

[4] https://github.com/Yquetzal/spaceCorrectedLouvainDC.

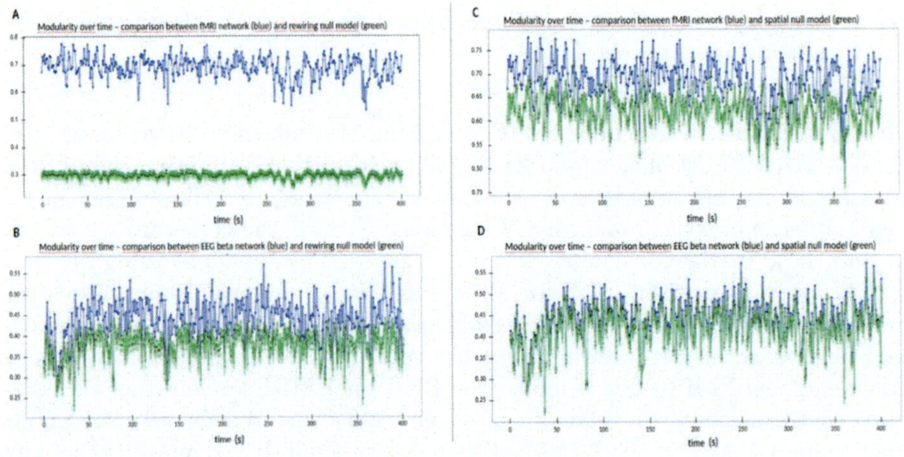

Fig. 1. Temporal variation of modularity (blue) for fMRI (A, C) and EEG beta (B, D), in comparison to the rewiring null model (green, A, B) and to the degree constrained spatial null model (C, D), for arbitrary subject. (Color figure online)

Table 1. Percentage of time points for which both modalities functional networks reflect a clustered structure in comparison to both rewiring and degree constrained spatial null model - for each frequency band, averaged for all subjects.

Percentage (%)	delta	theta	alpha	beta	gamma
Rewiring	60.6±6.5	56.8±5.3	59.1±5.3	59.5±7.1	65.3±11.4
Spatial	12.9±3.5	7.0±2.0	9.1±1.2	7.2±1.9	9.9±4.0

to oscillate between segregated and integrated states [4,5,15]. Besides this, there is a significant difference between fMRI and EEG functional networks, as the first one appears to possess a more clustered structure. This resulted in the selection of time points for both modalities being almost entirely constrained by the EEG. Furthermore, the degree constrained spatial null model presents a somewhat clustered topology, suggesting an important contribution of the spatial constraints for the structure observed. This is not surprising, since it has been reported a general tendency for the clusters, in functional networks, to be composed by regions that are near one another [3]. In particular, the spatial null model's global metrics appear to be almost identical to the EEG functional networks', suggesting that the topology detected for this modality might be a result of the spatial constraints imposed. This points to a higher susceptibility of the EEG's connectivity to these proximity constraints, which might result from its intrinsic lower spatial resolution as well as to its low signal-to-noise ratio (SNR). Nevertheless, there was still statistically significant structure being detected for some time points, specially for the delta frequency band, supposedly due to its ability to capture synchronized oscillations between brain regions at a longer distance than higher frequencies [13,22].

Table 2. Median modularity values associated to the communities extracted for fMRI and EEG frequency bands over time using the Louvain algorithm, averaged for all subjects. These values were computed considering each set of selected time points, using the rewiring null model, for every EEG-fMRI pair, for all frequency bands.

Modularity (Q)	delta	theta	alpha	beta	gamma
Q_{EEG}	0.455±0.004	0.466±0.002	0.447±0.003	0.465±0.010	0.428±0.015
Q_{fMRI}	0.700±0.004	0.700±0.005	0.697±0.005	0.699±0.004	0.699±0.015

3.2 Louvain Algorithm

The overall results of the community analysis with the Louvain algorithm, using the time points selected with the rewiring null model, are reported hereafter, for both EEG and fMRI functional networks. Table 2 summarizes the median modularity obtained for each modality, averaging for all subjects. Fig. 2 represents the range of modularity and NMI values obtained from the comparison of the community structure of both functional networks. Figure 3 shows the correlation over time between the two modalities, for arbitrary frequency band and subject.

In line with the previous observations, the fMRI functional networks show a more modular configuration than the EEG, which is also in accordance with previous dFC studies [33]. The less modular topology retrieved for the EEG might be due: i) to a worse quality of the data collected, as it is more affected by artifacts [30]; ii) to lack of sensibility of the technique to capture the topology of the underlying functional networks [23]; or iii) due to the difficulty in performing an accurate source reconstruction, specially for resting-state data [12].

Regarding the comparison between the two modalities' captured topology, it was found a low-to-moderate similarity, as it can be observed from the NMI results, which is in line with previous reports comparing EEG and fMRI static connectomes [35] for the same dataset. These results are also in accordance to studies examining dFC with both modalities, reporting a link between the two [1,8,13,34]. Nevertheless, this similarity is not particularly high, which might be due to the lack of modular topology for the EEG networks, as discussed. It might also be that this modality captures different interactions, leading to a more integrated topology instead of the segregated one found for the fMRI networks. In fact, it has been shown in [26] that EEG functional connectivity clusters into groups of brain regions differently than the fMRI functional connectivity and that these clusters appear to be extended in space, with a lower connectivity within modules than between them. Moreover, from the coloured NMI arrays, it is noticeable an oscillation in similarity over time, which was found to be specific to each frequency band. This not surprising considering past studies reporting a different contribution of each EEG frequency band to the BOLD connectivity dynamics [13,38], that varies across space [36], with a more local topology captured for higher frequency bands, such as the gamma band, and a more global connectivity for lower ones, like the delta band [22,34].

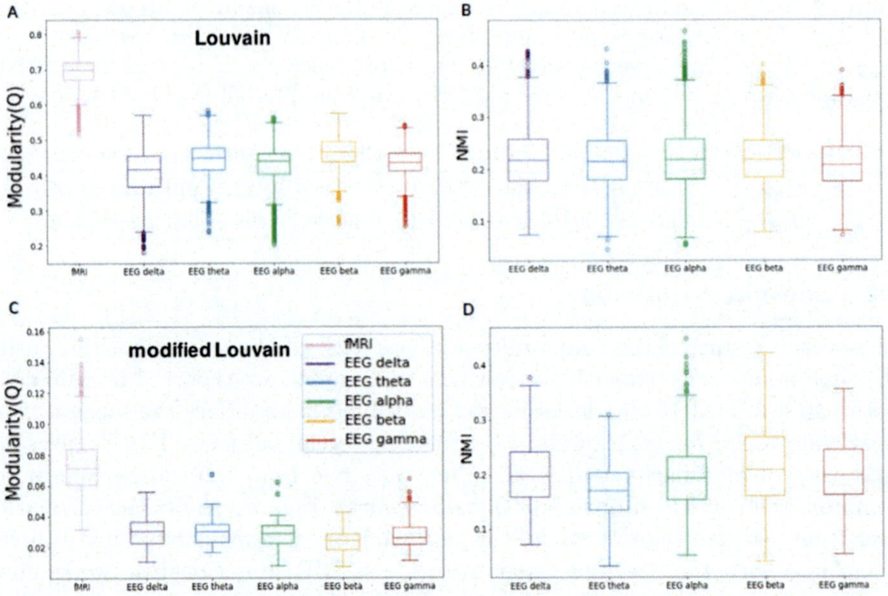

Fig. 2. Range of modularity for both EEG and fMRI networks (A, C), for all frequency bands, as well as range of NMI values (B, D) regarding the comparison of both modalities' communities obtained over time, with the Louvain algorithm (A, B) and with modified Louvain algorithm (C, D), for all subjects.

3.3 Modified Louvain Algorithm

To further analyse the spatiotemporal organization and contemplate its spatial embedding, the community analysis was performed with the modified Louvain algorithm, using the time points selected with the spatial null model. Table 3 summarizes the median modularity obtained for each modality, averaging for all subjects. Figure 2 shows the similarity between EEG and fMRI networks still arises beyond the influence of space, for all frequency bands, while Fig. 4 shows the comparison between the communities obtained over time with and without the spatial constraints, for arbitrary frequency band and subject.

Spatial constraints seem to explain the majority of the topology observed in EEG and fMRI functional networks, as observed in Sect. 3.1. Nevertheless, it was still possible to retrieve some significant community structure beyond these expectations, despite being associated to quite low modularity values (Table 3). When comparing the communities obtained with the regular and modified version of the Louvain algorithm, it was found an overall high similarity, but not a complete match. This points to the existence of relevant spatial patterns that arise out of functional necessity and not just as a consequence of space, even if not to a great extent. Furthermore, these similarity values are lower for the EEG, implying that, on top of not having a clear modular structure as the fMRI networks, the spatial effects have a higher impact in EEG networks' topology.

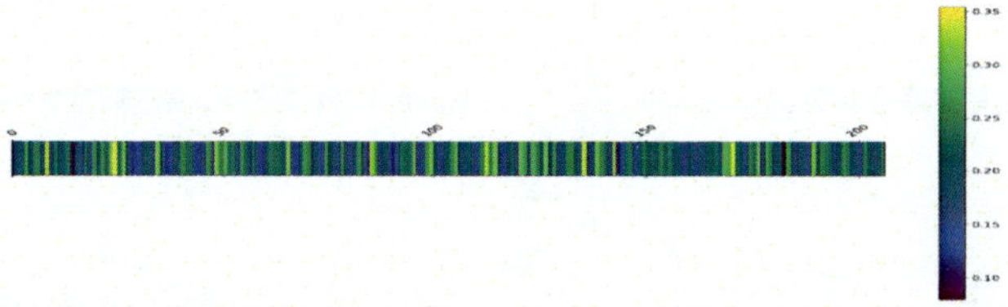

Fig. 3. NMI coloured array regarding the comparison of the communities obtained over time with the Louvain algorithm, between fMRI and EEG alpha band, for arbitrary subject. (Color figure online)

Table 3. Median modularity values associated to the communities extracted for fMRI and EEG frequency bands over time using the modified Louvain algorithm, averaged for all subjects. These values were computed considering each set of selected time points, using the degree constrained spatial null model, for every EEG-fMRI pair, for all frequency bands.

Modularity (Q)	delta	theta	alpha	beta	gamma
Q_{EEG}	0.036±0.003	0.034±0.002	0.034±0.001	0.032±0.002	0.033±0.002
Q_{fMRI}	0.072± 0.002	0.071± 0.005	0.073± 0.003	0.072± 0.004	0.071± 0.003

Comparing the two modalities community structure beyond the spatial constraints, an overall lower similarity was obtained, in comparison to the one presented in Sect. 3.2 (see Fig. 2). This suggests that part of the similarity between the two is guaranteed by the underlying spatial embedding. Even so, it was still retrieved a partially similar modular configuration beyond that, which supports a link between the EEG and fMRI dFC. However, it is important to take into consideration that this analysis was done only for the few time points deviating from the spatial null model (around 9%).

3.4 Multiplex Louvain Algorithm

Since it was not found a total match between the topology captured by two modalities on a global level, one can speculate that these complementary techniques capture different information regarding the underlying neuronal activity and its functional organization. The improved median modularity resulting from the multiplex Louvain algorithm analysis is reported in Table 4, as well as the individual values obtained for each modality.

One can immediately notice that the individual modularity values obtained are lower than the single-layer ones reported in Sect. 3.2. Meaning that the multi-layer approach finds clusters common to both modalities that were not captured previously, as these partitions possessed too low modularity to be selected by the community detection procedure. This suggests that using EEG and fMRI together allows the capture of modules of synchronised activity that otherwise would not be found if looking at each functional network individually. These findings are in line with two previous studies that performed a joint-analysis of

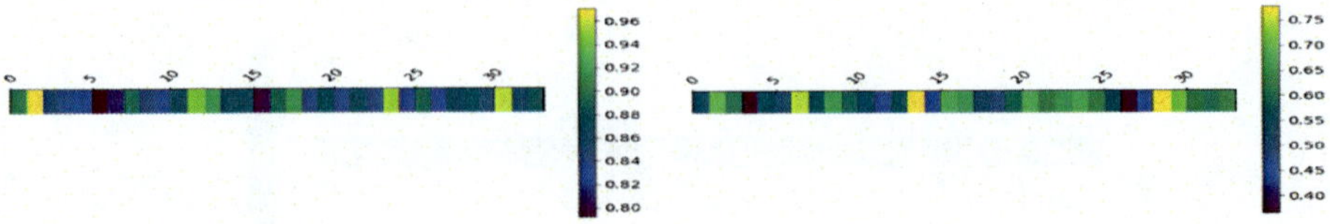

Fig. 4. NMI coloured arrays regarding the comparison of the communities obtained over time with and without the spatial constraints, between fMRI and EEG theta band, respectively, for arbitrary subject. (Color figure online)

Table 4. Median modularity (Q) values associated to the common communities extracted from both fMRI and EEG frequency bands over time using the multiplex Louvain algorithm, averaged for all subjects. These values were computed considering each set of selected time points, using the rewiring null model, for every EEG-fMRI pair, for all frequency bands.

Modularity (Q)	delta	theta	alpha	beta	gamma
$Q_{multiplex}$	0.747±0.004	0.752±0.004	0.748±0.005	0.755±0.005	0.751±0.005
Q_{EEG}	0.091±0.009	0.123±0.009	0.105±0.005	0.148±0.011	0.102±0.014
Q_{fMRI}	0619±0.011	0.591±0.010	0.603±0.009	0.571±0.009	0.615±0.007

these modalities, by means of a hybrid independent component analysis [33] and by building a multimodal graph, joining the EEG and fMRI nodes into a single network [38] to identify new connectivity structure. Furthermore, the modularity found was statistically significant in comparison to a multiplex spatial null model for most time points, implying that it is not just the spatial embedding that leads to the common partitions found.

4 Conclusions

From this work it is possible to draw several conclusions. First of all, the EEG and fMRI functional connectivity seem to capture different information on a global level. FMRI networks showed more modular configuration, consistent over time, while EEG ones captured a less clustered topology, with each frequency band capturing a slightly different structure oscillating across time. Moreover, when combining the two modalities, significant communities were extracted that would not be captured otherwise. Secondly, both functional networks' organization is mostly explained by the spatial embedding, giving preference to close connections. Nevertheless, relevant communities were still obtained beyond those constraints, for both fMRI and EEG, in particular for the delta, alpha and gamma bands. Finally, despite the differences reported, there is a similarity between the modalities' topology over time, again mostly explained by the spatial embedding. Nonetheless, when regressing out the influence of space, a small similarity was still retrieved for a set of time points. Therefore, it is possible to conclude that, even though fMRI and EEG functional connectomes are slightly linked, the two essentially capture different information on a topological level. As such, combining the modalities seems desirable to characterize the brain's complex activity and to distinguish different states and conditions. Furthermore, this work reinforces the importance of analysing functional networks choosing appropriate null models to retrieve truly meaningful features.

Acknowledgments. This work was supported by national funds through FCT – Fundação para a Ciência e Tecnologia, under grant UIDB/50021/2020 and UIDP/50009/2020. AST acknowledges support by the FCT through the LASIGE Research Unit, ref. UIDB/00408/2020 and ref. UIDP/ 00408/2020. APF acknowledges support by FCT through ref. UIDB/50021/2020. PF acknowledges support by FCT through LARSyS, ref. UDIP/50009/2020.

References

1. Abreu, R., Jorge, J., Leal, A., Koenig, T., Figueiredo, P.: EEG microstates predict concurrent fMRI dynamic functional connectivity states. Brain Topogr. **34**(1), 41–55 (2020). https://doi.org/10.1007/s10548-020-00805-1
2. Baillet, S., Mosher, J.C., Leahy, R.M.: Electromagnetic brain mapping. IEEE Signal Process. Mag. **18**(6), 14–30 (2001). https://doi.org/10.1109/79.962275
3. Bassett, D.S., Stiso, J.: Spatial brain networks. C R Phys. **19**(4), 253–264 (2018). https://doi.org/10.1016/j.crhy.2018.09.006
4. Betzel, R.F., Byrge, L., Esfahlani, F.Z., Kennedy, D.P.: Temporal fluctuations in the brain's modular architecture during movie-watching. Neuroimage **213**, 116687 (2020). https://doi.org/10.1016/j.neuroimage.2020.116687
5. Betzel, R.F., Fukushima, M., He, Y., Zuo, X.N., Sporns, O.: Dynamic fluctuations coincide with periods of high and low modularity in resting-state functional brain networks. Neuroimage **127**, 287–297 (2016). https://doi.org/10.1016/j.neuroimage.2015.12.001
6. Blondel, V.D., Guillaume, J.L., Lambiotte, R., Lefebvre, E.: Fast unfolding of communities in large networks. J. Stat. Mech: Theory Exp. **2008**(10), P10008 (2008). https://doi.org/10.1088/1742-5468/2008/10/P10008
7. Bordier, C., Nicolini, C., Bifone, A.: Graph analysis and modularity of brain functional connectivity networks: Searching for the optimal threshold. Front. Neurosci. **11**, 441 (2017). https://doi.org/10.3389/fnins.2017.00441
8. Bréchet, L., Brunet, D., Birot, G., Gruetter, R., Michel, C.M., Jorge, J.: Capturing the spatiotemporal dynamics of self-generated, task-initiated thoughts with EEG and fMRI. Neuroimage **194**, 82–92 (2019). https://doi.org/10.1016/j.neuroimage.2019.03.029
9. Bullmore, E., Sporns, O.: Complex brain networks: Graph theoretical analysis of structural and functional systems. Nat. Rev. Neurosci. **10**, 186–198 (2009). https://doi.org/10.1038/nrn2575
10. Cabral, J., et al.: Cognitive performance in healthy older adults relates to spontaneous switching between states of functional connectivity during rest. Sci. Rep. **7**, 5135 (2017). https://doi.org/10.1038/s41598-017-05425-7
11. Cazabet, R., Borgnat, P., Jensen, P.: Enhancing space-aware community detection using degree constrained spatial null model. In: Gonçalves, B., Menezes, R., Sinatra, R., Zlatic, V. (eds.) CompleNet 2017. SPC, pp. 47–55. Springer, Cham (2017). https://doi.org/10.1007/978-3-319-54241-6_4

12. Custo, A., Van De Ville, D., Wells, W.M., Tomescu, M.I., Brunet, D., Michel, C.M.: Electroencephalographic Resting-State Networks: Source Localization of Microstates. Brain Connect. **7**(10), 671–682 (2017). https://doi.org/10.1089/brain.2016.0476
13. Deligianni, F., Centeno, M., Carmichael, D.W., Clayden, J.D.: Relating resting-state fMRI and EEG whole-brain connectomes across frequency bands. Front. Neurosci. **8**(258) (2014). https://doi.org/10.3389/fnins.2014.00258
14. Desikan, R.S., et al.: An automated labeling system for subdividing the human cerebral cortex on MRI scans into Gyral based regions of interest. Neuroimage **31**(3), 968–980 (2006). https://doi.org/10.1016/j.neuroimage.2006.01.021
15. Dimitriadis, S., Laskaris, N., Tsirka, V., Vourkas, M., Sifis, M.: An EEG study of brain connectivity dynamics at the resting state. Nonlinear Dyn. Psychol. Life Sci. **16**(1), 5–22 (2012)
16. Esfahlani, F.Z., Bertolero, M.A., Bassett, D.S., Betzel, R.F.: Space-independent community and hub structure of functional brain networks. Neuroimage **211**, 116612 (2020). https://doi.org/10.1016/j.neuroimage.2020.116612
17. Farahibozorg, S.R., Henson, R.N., Hauk, O.: Adaptive cortical parcellations for source reconstructed EEG/MEG connectomes. Neuroimage **169**, 23–45 (2018). https://doi.org/10.1016/j.neuroimage.2017.09.009
18. Fukushima, M., Sporns, O.: Comparison of fluctuations in global network topology of modeled and empirical brain functional connectivity. PLoS Comput. Biol. **14**(9), e1006497 (2018). https://doi.org/10.1371/journal.pcbi.1006497
19. van den Heuvel, M.P., Hulshoff Pol, H.E.: Exploring the brain network: a review on resting-state fMRI functional connectivity. Eur. Neuropsychopharmacol. **20**(8), 519–534 (2010). https://doi.org/10.1016/j.euroneuro.2010.03.008
20. Jorge, J., Bouloc, C., Bréchet, L., Michel, C.M., Gruetter, R.: Investigating the variability of cardiac pulse artifacts across heartbeats in simultaneous EEG-fMRI recordings: a. Neuroimage **191**, 21–35 (2019). https://doi.org/10.1016/j.neuroimage.2019.02.021
21. Lewin, J.S.: Functional MRI: an introduction to methods. J. Magn. Reson. Imaging **17**(3), 383–383 (2003). https://doi.org/10.1002/jmri.10284
22. Lopes da Silva, F.: EEG and MEG: relevance to neuroscience. Neuron **80**(5), 1112–1128 (2013). https://doi.org/10.1016/j.neuron.2013.10.017
23. Mahjoory, K., Nikulin, V.V., Botrel, L., Linkenkaer-Hansen, K., Fato, M.M., Haufe, S.: Consistency of EEG source localization and connectivity estimates. Neuroimage **152**, 590–601 (2017). https://doi.org/10.1016/j.neuroimage.2017.02.076
24. Mele, G., Cavaliere, C., Alfano, V., Orsini, M., Salvatore, M., Aiello, M.: Simultaneous EEG-fMRI for functional neurological assessment. Front. Neurol. **10** (2019). https://doi.org/10.3389/fneur.2019.00848
25. Mucha, P.J., Richardson, T., Macon, K., Porter, M.A., Onnela, J.P.: Community structure in time-dependent, multiscale, and multiplex networks. Science **328**(5980), 876–878 (2010). https://doi.org/10.1126/science.1184819
26. Nentwich, M., et al.: Functional connectivity of EEG is subject-specific, associated with phenotype, and different from fMRI. Neuroimage **218**, 117001 (2020). https://doi.org/10.1016/j.neuroimage.2020.117001
27. Nolte, G., Bai, O., Wheaton, L., Mari, Z., Vorbach, S., Hallett, M.: Identifying true brain interaction from EEG data using the imaginary part of coherency. Clin. Neurophysiol. **115**(10), 2292–2307 (2004). https://doi.org/10.1016/j.clinph.2004.04.029

28. Poldrack, R.A., Nichols, T., Mumford, J.: Handbook of Functional MRI Data Analysis. Cambridge University Press (2011). https://doi.org/10.1017/cbo9780511895029
29. Preti, M.G., Bolton, T.A., Van De Ville, D.: The dynamic functional connectome: state-of-the-art and perspectives. Neuroimage **160**, 41–54 (2017). https://doi.org/10.1016/j.neuroimage.2016.12.061
30. Puxeddu, M.G., Petti, M., Pichiorri, F., Cincotti, F., Mattia, D., Astolfi, L.: Community detection: comparison among clustering algorithms and application to EEG-based brain networks. In: Proceedings of the Annual International Conference of the IEEE Engineering in Medicine and Biology Society, EMBS, pp. 3965–3968 (2017). https://doi.org/10.1109/EMBC.2017.8037724
31. Roberts, J.A., et al.: The contribution of geometry to the human connectome. NeuroImage **124**(PtA), 379–393 (2016). https://doi.org/10.1016/j.neuroimage.2015.09.009
32. Samu, D., Seth, A.K., Nowotny, T.: Influence of wiring cost on the large-scale architecture of human cortical connectivity. PLoS Comput. Biol. **10**(4), e1003557 (2014). https://doi.org/10.1371/journal.pcbi.1003557
33. Wirsich, J., Amico, E., Giraud, A.L., Goñi, J., Sadaghiani, S.: Multi-timescale hybrid components of the functional brain connectome: A bimodal EEG-fMRI decomposition. Network Neurosci. **4**(3), 658–677 (2020). https://doi.org/10.1162/netn_a_00135
34. Wirsich, J., Giraud, A.L., Sadaghiani, S.: Concurrent EEG- and fMRI-derived functional connectomes exhibit linked dynamics. Neuroimage **219**, 116998 (2020). https://doi.org/10.1016/j.neuroimage.2020.116998
35. Wirsich, J., et al.: The relationship between EEG and fMRI connectomes is reproducible across simultaneous EEG-fMRI studies from 1.5t to 7t. NeuroImage **231**, 117864 (2021). https://doi.org/10.1016/j.neuroimage.2021.117864
36. Wirsich, J.: Complementary contributions of concurrent EEG and fMRI connectivity for predicting structural connectivity. Neuroimage **161**, 251–260 (2017). https://doi.org/10.1016/j.neuroimage.2017.08.055
37. Wu, G.R., Liao, W., Stramaglia, S., Ding, J.R., Chen, H., Marinazzo, D.: A blind deconvolution approach to recover effective connectivity brain networks from resting state fMRI data. Med. Image Anal. **17**(3), 365–374 (2013). https://doi.org/10.1016/j.media.2013.01.003
38. Yu, Q., et al.: Building an EEG-fMRI multi-modal brain graph: a concurrent EEG-fMRI study. Front. Hum. Neurosci. **10**, 476 (2016). https://doi.org/10.3389/fnhum.2016.00476

Comparative Study of Random Walks with One-Step Memory on Complex Networks

Miroslav Mirchev(✉), Lasko Basnarkov, and Igor Mishkovski

Faculty of Computer Science and Engineering, Ss. Cyril and Methodius University in Skopje, Rudjer Boshkovikj 16, 1000 Skopje, North Macedonia
{miroslav.mirchev,lasko.basnarkov,igor.mishkovski}@finki.ukim.mk

Abstract. We investigate searching efficiency of different kinds of random walk on complex networks which rely on local information and one-step memory. For the studied navigation strategies we obtained theoretical and numerical values for the graph mean first passage times as an indicator for the searching efficiency. The experiments with generated and real networks show that biasing based on inverse degree, persistence and local two-hop paths can lead to smaller searching times. Moreover, these biasing approaches can be combined to achieve a more robust random search strategy. Our findings can be applied in the modeling and solution of various real-world problems.

Keywords: Random walk · Complex network · Graph · Graph search

1 Introduction

Random walk is a ubiquitous concept that describes wandering in certain space in which the location where the walker will be in the next moment is chosen randomly. In complex networks it can applied for modeling diverse phenomena like searching through information networks [1], diffusion of information, ideas and viruses in social networks, stock market fluctuations, and solving various problems such as page ranking in the web [19], semi-supervised graph labeling [10,29], link prediction in graphs [2], and graph representation learning [13,17].

Since the onset of interest in complex networks, various models of random walk on top of them have been proposed. The standard uniform random walk is based on randomly choosing the next node in the walk with equal probability from all neighbors of the node where the walker currently is. By applying master equation approach [18] or Markov chain theory [12] one can obtain theoretical results for a key quantity in the random walk – the mean first passage time (MFPT), that represents the expected number of steps needed for the walker to reach randomly chosen target for the first time. Using the same formalism, various modifications of the uniform random walk have been applied that exploit the local properties of the network, aimed at improving the search time. One approach is based on the degrees of the neighbors [9], particularly when biasing proportionally to the inverse degree of the next node [4,6]. Some authors have

A. S. Teixeira et al. (Eds.): CompleNet 2023, SPCOM, pp. 14–25, 2023.
https://doi.org/10.1007/978-3-031-28276-8_2

considered local neighborhood exploration by random walks using marking as well as biasing based on neighbors degrees [5]. In another approach memory is applied where the probability to jump to some next node depends on the current, but also on the previously visited one [3,4,7]. Other problems that have recently received attention are random walk on networks with resetting [21], multiple simultaneous random walks [20], and random walk on hypergraphs [8].

The theoretical expressions for calculating MFPT in random walks with one-step memory presented in [4] provide a useful testbed that can be employed for comparing various biasing strategies in relatively small networks. Nevertheless, the findings can be then applied to networks with arbitrary sizes. In this work, we aim to study and combine different approaches with local information in order to see whether further improvement is possible. We study five types of random walks with one-step memory: simple forward going, inverse degree biased, two-hop paths based, persistent, and we introduce a combination of persistent and inverse degree biased. For comparison in our study we also include two standard random walks without memory: uniform and inverse degree biased. Our findings can be applied for potential improvements in the study of a wide range of problems mentioned at the beginning of this introduction.

In Sect. 2 we describe the theoretical expressions for calculating MFPTs in random walks with one-step memory on complex networks represented as graphs. Several graph searching strategies using such random walks are described in Sect. 3. In Sect. 4 we present the results obtained with the theoretical expressions and numerical simulations on several generated and real complex networks, while in Sect. 5 we give some general conclusions.

2 Mean First Passage Time of Random Walks with One-Step Memory on Complex Networks

In this section we briefly restate the main analytical results from [4] for representing a random walk with a one-step memory over complex networks, but a detailed explanation of the expressions derivation can be found in the original paper. For the sake of simplicity we use notations for undirected networks, although the same theory also holds for directed networks. A complex network given as a graph $G(V, E)$ composed of a set of vertices V, $|V| = N$, and a set of edges E, $|E| = L$, can be represented by an adjacency matrix $\mathbf{A}_{N\times N}$. We study discrete-time random walks with a one-step memory, so that a random walk that in the previous steps has visited nodes $\{\dots, o, p, q, r\}$ and currently is in node s, can visit a next node t with a probability

$$p(t|s, r, q, p, o, \dots) = p(t|s, r). \tag{1}$$

In order to represent such a random walk with a Markov chain instead of using the nodes as states we use the links between the nodes. The transition matrix of the corresponding Markov chain $\mathbf{P}_{L\times L}$ will have elements $p_{rs,st} = p(t|s, r)$, $\forall rs, st, \in E$. These elements can take arbitrary values that represent probabilities, depending on the chosen random walk.

The random walk can be initialized by starting from node a and then passing to a random neighbor b, so from that moment on the transitions can be made according to $\mathbf{P}$. The problem of finding a target node z can be represented as reaching any state yz, where y is any neighbor of z. This process can be represented by an absorbing Markov chain with a transition matrix $\mathbf{P}_{(z)}$ where all states yz are absorbing, while all other transitions states $ij, j \neq z$ are transient. The theory of absorbing Markov chains and particularly the mean time to absorption (MTA) can be then used to calculate the mean first passage time (MFPT) from a to z [12,24]. For simplicity we assume that the random walk never starts from the target z, so for the MFPT calculation we can safely omit all states $zy, \forall y$. The transition matrix $\mathbf{P}_{(z)}$ takes the form

$$\mathbf{P}_{(z)} = \begin{vmatrix} \mathbf{Q}_{(z)} & \mathbf{R}_{(z)} \\ \mathbf{0} & \mathbf{I} \end{vmatrix}, \tag{2}$$

where $\mathbf{Q}_{(z)}$ is an $(L-k_z) \times (L-k_z)$ matrix containing the transition probabilities among transient states, $\mathbf{R}_{(z)}$ is an $(L-k_z) \times k_z$ matrix representing the transitions from the transient to the absorbing states, and $\mathbf{I}$ is an $k_z \times k_z$ identity matrix. The fundamental matrix for the corresponding Markov chain contains the expected number of steps that a random walk starting from any transient state ab is present in another transient state ij can be expressed as the infinite sum

$$\mathbf{Y}_{(z)} = \mathbf{I} + \mathbf{Q}_{(z)} + \mathbf{Q}_{(z)}^2 + \cdots. \tag{3}$$

The powers of $\mathbf{Q}_{(z)}$ diminish as $n \to \infty$, and $\mathbf{Y}_{(z)}$ converges towards

$$\mathbf{Y}_{(z)} = \left(\mathbf{I} - \mathbf{Q}_{(z)}\right)^{-1}. \tag{4}$$

Then we can obtain a vector containing all the MTA from all possible initial states ab by multiplying with a vector of ones $\mathbf{1}$

$$\mu_{(z)} = \mathbf{Y}_{(z)}\mathbf{1}. \tag{5}$$

Then the MFPT from a to z can be calculated as [4]

$$m_{a,z} = 1 + \frac{1}{k_a} \sum_{b \in \mathcal{N}_a} \mu_{(z),ab}. \tag{6}$$

A Global Mean First Passage Time (GMFPT) [26] can be found by averaging over all starting nodes a as

$$g_z = \frac{1}{N-1} \sum_{\substack{a=1 \\ a \neq z}}^{N} m_{a,z}. \tag{7}$$

By repeating the same procedure and averaging over all target nodes z we can express a Graph MFPT (GrMFPT) as [6]

$$G = \sum_{z=1}^{N} g_z, \tag{8}$$

which will be used in the rest of the paper for comparing several different strategies of random walks with one-step memory.

3 Graph Search Algorithms Based on Random Walks

In this section we describe several different strategies for graph search using random walks with memory. We also include two classical random walks without memory: a uniform random walk and an inverse degree biased random walk. In a previous work [4], we considered the application for a random walk searching strategy based on the number of two-hop paths towards the next node in the walk, which we call "two-hop random walk with memory". However, the results showed that in directed complex networks this strategy does not brings improvements and simply choosing the next node solely based on its inverse degree resulted in shorter hitting times. Therefore, in this paper we also consider four other random walk strategies with one-step memory. The first one simply avoids going back, which was thoroughly studied in [7], and we refer it as "forward random walk with memory". The second strategy, which we call "inverse degree random walk with memory", in addition to avoiding going backwards chooses the next node based on its degree. Another approach called "persistent random walk with memory" which employs biasing towards more distant nodes by avoiding neighbors of the previously visited node, was numerically studied in [3], but here we further provide calculations based on the theoretical expressions. Moreover, we examine a hybrid of the persistent and the inverse degree random walks with memory, in order to combine their strengths and help cover their weaknesses. For calculating MFPTs and GrMFPT in the random walks with memory we will use the theoretical expressions from [4], described in the previous section, for all networks in our study. However, in the random walks without memory we will use the expressions presented in [4] for finding GrMFPT in the generated networks, and standard expressions based on absorbing Markov chains [24] for the real networks due to numerical problems with the expressions from [4].

3.1 Classical Random Walks Without Memory

Uniform Random Walk (U-RW) - at each step the random walk makes a transition from node s to any of its neighbors t with an equal probability $p_{st} = 1/k_s$.

Inverse Degree Random Walk (ID-RW) - the visiting probability of s to a neighboring node t is inversely proportional to its node degree $1/k_t$, hence

$$p_{st} = \frac{1/k_t}{\sum_{t \in \mathcal{N}_s} 1/k_t}. \tag{9}$$

3.2 Random Walks with Memory

In the random walks with memory, we denote the previous visited node as r, the current node as s, and the potential next nodes as t.

Forward Random Walk with Memory (F-RWM) - the random walk avoids going back, by exploiting the one-step memory, unless there is no way to keep going forward. The probability of visiting the other neighboring nodes is equal and expressed as

$$p_{rs,st} = 1/(k_s - 1), \forall r \neq t, \tag{10}$$

while the probability of going back can be written as

$$p_{rs,sr} = \begin{cases} 1 & \text{if r is the only neighbor of s,} \\ 0 & \text{otherwise.} \end{cases} \tag{11}$$

Inverse Degree Random Walk with Memory (ID-RWM) - if a random walk has previously visited r and currently is in s would visit some of its other neighbors with a probability.

$$p_{rs,st} = \frac{1/k_t}{\sum_{t \in \mathcal{N}_s \setminus \{r\}} 1/k_t}, \tag{12}$$

while it avoids going back to r by following Eq. (11).

Two-hops Random Walk with Memory (2H-RWM) - the probabilities are proportional to the number of two-hop paths that lead toward the target nodes, so the visiting probabilities are given as

$$p_{rs,st} = \frac{\frac{1}{b_{rt}}}{\sum_{u \in \mathcal{N}_s} \frac{1}{b_{ru}}}, \tag{13}$$

where b_{rt} is the number of two-hop paths between r and t, or the elements of $\mathbf{B} = \mathbf{A}^2$.

Persistent Random Walk with Memory (P-RWM) - the random walk avoids going backwards or towards the neighbours of the previously visited node. Let us denote with $N_1 = |\mathcal{N}_r \cap \mathcal{N}_s|$ the number of common neighbors of s with r which it visits each with a probability p_1, and let $N_2 = |\mathcal{N}_s \setminus \{\mathcal{N}_r \cap \mathcal{N}_s\}|$ be the number of other neighbors, which it visits with a probability p_2 each, and let p_0 be a probability of immediately going back to r. We can then write $p_0 + N_1 p_1 + N_2 p_2 = 1$. By introducing the parameters $p_2/p_1 = \alpha$ and $p_0/p_1 = \beta$, the previous expression can be rewritten as $p_1(\beta + N_1 + \alpha N_2) = 1$. Hence, the probability of going back to r is given by

$$p_{rs,st} = \begin{cases} p_0 = \frac{\beta}{C} & \text{if } r = t, \\ p_1 = \frac{1}{C} & \text{if } t \in \mathcal{N}_r \cap \mathcal{N}_s, \\ p_2 = \frac{\alpha}{C} & \text{if } t \in \mathcal{N}_s \setminus \{\mathcal{N}_r \cap \mathcal{N}_s\}, \end{cases} \tag{14}$$

where $C = \beta + N_1 + \alpha N_2$ is a normalization coefficient. A detailed analysis of the effects of the parameters α and β on GrMFPT can be found in [3].

Persistent Inverse Degree Random Walk with Memory (PID-RWM) - combines the strengths of the persistent and inverse degree biased random walks. We have excluded the possibility for going back to r so $p_{rs,sr} = 0$, except when it is the only option $p_{rs,sr} = 1$. The probability of going toward a common neighbor with r will become

$$p_{rs,st} = \begin{cases} \frac{1/k_t}{C} & \text{if } t \neq r, t \in \mathcal{N}_r \cap \mathcal{N}_s, \\ \frac{\alpha/k_t}{C} & \text{if } t \neq r, t \in \mathcal{N}_s \setminus \{\mathcal{N}_r \cap \mathcal{N}_s\}, \end{cases} \tag{15}$$

where the normalization coefficient is

$$C = \sum_{t \in \mathcal{N}_r \cap \mathcal{N}_s} 1/k_t + \sum_{t \in \mathcal{N}_s \setminus \{\mathcal{N}_r \cap \mathcal{N}_s\}} \alpha/k_t. \tag{16}$$

Once again the random walk avoids going back to r by following Eq. (11). We like to note that for $\alpha = 1$, PID-RWM becomes identical to ID-RWM.

4 Results

First, we examine the GrMFPT for five types of random walks: classical, inverse degree, two-hop with memory, inverse degree with memory and forward with memory, for three complex networks models: Barabási-Albert (BA), Watts-Strogatz (WS), and Erdős-Rényi (ER) with undirected and directed links. In Fig. 1 we show the results calculated using the theoretical expressions and by numerical simulations of the random walks transitions. The results are averaged over 10 different network instances generated with the same parameter values, while the numerical simulations are further averaged across 10 repetitions of all node pairs. The rewiring probability in the WS model is $p_{\text{rew}} = 0.2$. In P-RWM and PID-WM, $\alpha = 10$ and $\beta = 0.01$, but one can further analyse the effects of varying α. As can be seen in Fig. 1a, for BA networks the ID-RWM and PID-RWM significantly outperform the other methods, and the difference is large for small $\langle k \rangle$ particularly with the similar ID-RW, which also employs inverse degree biasing but lacks memory. The results for WS networks presented in Fig. 1b show how the persistent random walks exploit the memory, with PID-RWM slightly outperforming P-RWM. In ER networks (Fig. 1c), the inverse degree biasing proves crucial again and the ID-RWM and PID-RWM show best searching performance. As we have previously noted in [4], in directed ER networks (Fig. 1d) the simple inverse degree biasing without memory, still holds well with almost identical performance with the other inverse degree biased approaches ID-RWM and PID-RWM. These results can be expected as in sparse ER directed networks rarely ever two nodes are connected in both directions, hence, the going back avoidance is rarely exploited. In some real networks these bidirectional mutual connectivity can be present more often so the memory could be more beneficial there. Overall, we can conclude that the PID-RWM shows the best and most consistent performance across different network topologies and densities.

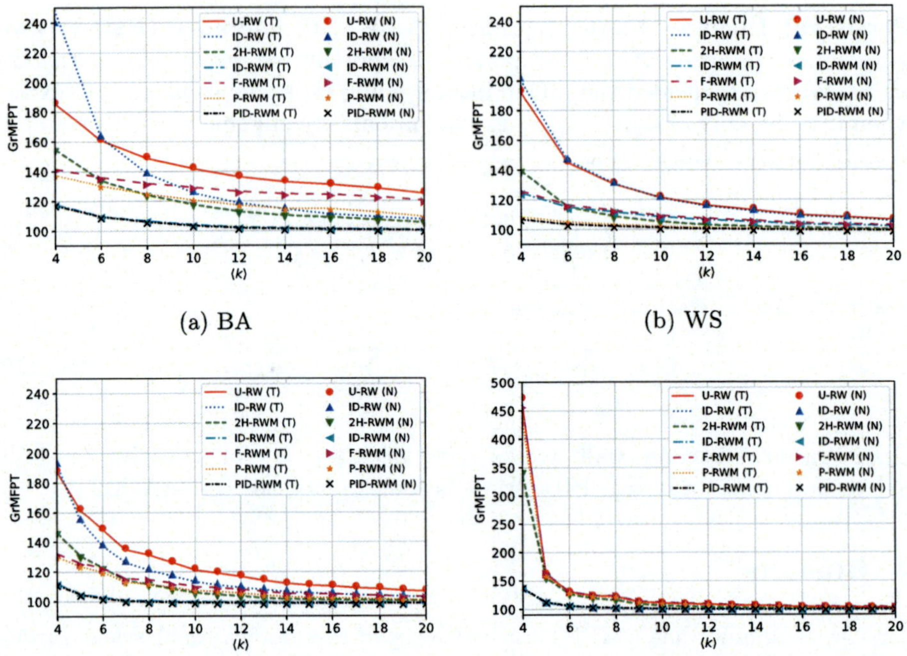

(a) BA (b) WS

(c) ER (d) ER directed

Fig. 1. GrMFPT in (a) BA, (b) WS, (c) ER, and (d) ER directed networks, with 100 nodes and varied average node degree $\langle k \rangle$ for 7 different random walks. The lines are theoretical values (T) and the markers numerical estimates (N).

We have also studied how the stationary distributions of the visiting probabilities are affected from the choice of the random walk strategy, by comparing them with a uniform distribution using the KL divergence

$$D_{\mathrm{KL}}(P||Q) = \sum_i P(i) \log \frac{P(i)}{Q(i)}, \tag{17}$$

and the results are shown in Fig. 2. As expected the ID-RW significantly equalizes the visiting probabilities, which is the reason behind the often observed shorter search times compared to the U-RW. In most cases ID-RW achives lowest D_{KL}, however, for ER networks with low $\langle k \rangle$ ID-RWM and PID-RWM achive slightly lower D_{KL}.

To gain a deeper insight into the effect of the rewired links in the WS model, we studied how the MFPT varies with the change of the rewiring probability p_{rew} in networks with $k = 4$ and $k = 6$ neighbors per node, and the results are given in Fig. 3. The WS model transitions from a regular lattice toward a completely random ER network, as p_{rew} is varied from 0 to 1. It can be seen how for small

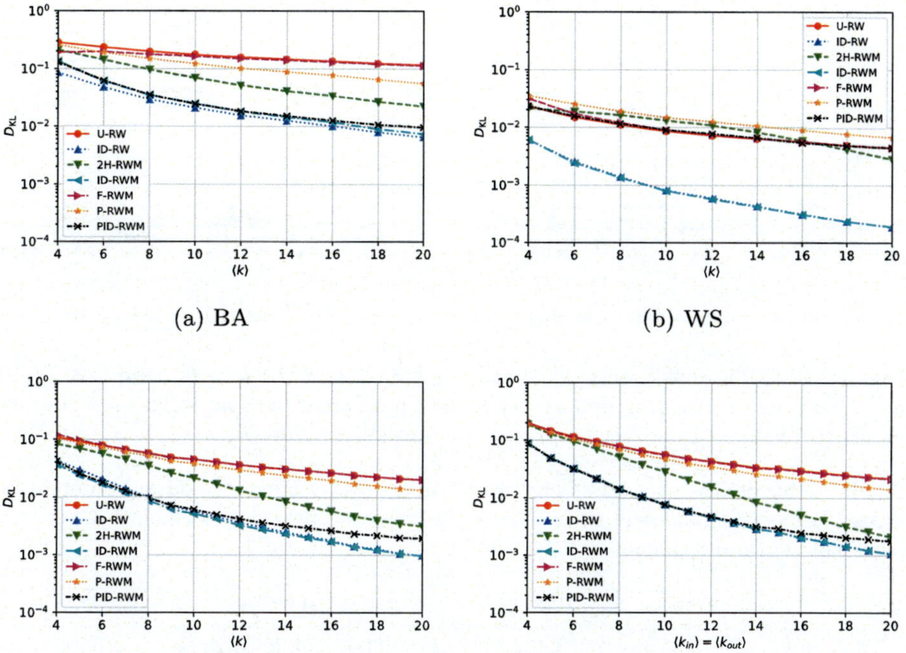

(a) BA (b) WS

(c) ER (d) ER directed

Fig. 2. Kullback-Leibler divergence of the stationary occupation probability of 7 different random walks from a uniform density in (a) BA, (b) WS, (c) ER, and (d) ER directed networks, with 100 nodes and varied average node degree $\langle k \rangle$.

p_{rew} PID-RWM and P-RWM behave similarly and have best performances, however, as p_{rew} increases the performance of P-RWM decreases eventually loosing the pace with ID-RWM, while PID-RWM keeps its performance at level with ID-RWM.

We also made comparison of the various random walks on several real networks and their main structural properties are given in Table 1. The first network is a representation of the Internet at level of autonomous systems derived from BGP logs [14], which is known to have the scale-free property. The second network is an excerpt from Wikipedia pages [27, 28], also having a scale-free property. This network was used in [27] to study human wayfinding to a given target through Wikipedia pages. The original dataset consists of 4592 nodes and 119882 links, but for our analysis we use the largest strongly connected component. The third network Euroroad is a representation of major European roads [25]. It is an undirected network and consists of 1174 nodes and 1417 edges, from which we take the largest connected component. The fourth network FB-Pages is a collection of Facebook pages and their mutual likes [23], and the fifth is a network of human diseases (Bio-diseasome) [11]. The sixth network CA-netscience depicts collaboration in publications between researchers in the field of network

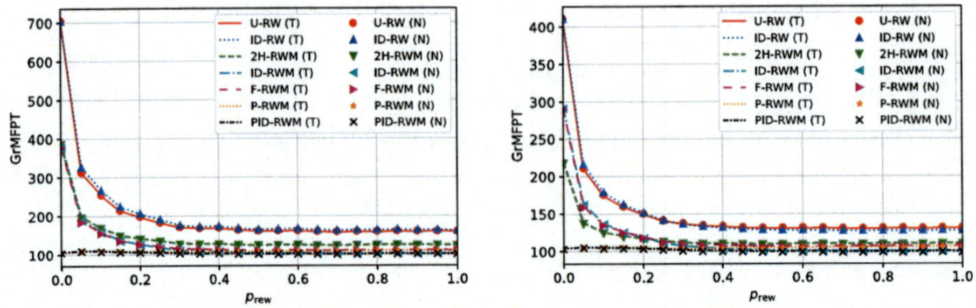

(a) WS model with $k = 4$

(b) WS model with $k = 6$

Fig. 3. GrMFPT in WS networks with (a) $k = 4$, and (b) $k = 6$, composed of 100 nodes with varied rewiring probability for seven different random walks. The lines are theoretical values (T) and the markers numerical estimates (N).

Table 1. Statistical properties of the real networks: number of nodes N, number of links L, density D, average node degree $\langle k \rangle$, average clustering coefficient C, average path length $\langle l \rangle$, and diameter d.

	Type	N	L	D	$\langle k \rangle$	C	$\langle l \rangle$	d
Internet	Undirected	6474	13233	0.0006	4.29	0.2522	3.7050	9
Wikipedia	Directed	4051	119000	0.0068	27.62	0.1892	3.1813	9
Euroroad	Undirected	1039	1305	0.0024	2.51	0.01890	18.3951	62
FB-Pages	Undirected	620	2102	0.0109	3.39	0.3309	5.0887	17
Bio-diseasome	Undirected	516	1188	0.0089	2.30	0.6358	6.501	15
CA-netscience	Undirected	379	914	0.0128	2.41	0.7412	6.042	17

science [16]. The first two datasets are taken from the SNAP dataset collection [15], and the last three from the Network repository [22]. A visualization of the last four networks is given in Fig. 4, using the Force atlas layout, where nodes with larger degree are colored darker.

The results with the real networks summarized in Table 2 show that the theoretical expressions are in accord with the numerical simulations. The first two networks are relatively larger, so for them we conducted only numerical simulations, while for the other networks we also provide results from the theoretical expressions. The numerical results for all networks are obtained by calculating the MFPT between 100000 randomly chosen node pairs. There were some numerical problems in the calculations of the GrMFPT for some networks using the analytical expressions for memoryless random walks from [4], so for them we used standard expressions based on absorbing Markov chains [24]. The results indicate that the random walks can behave differently in real networks, as their structure is not always very similar to networks generated with classical models.

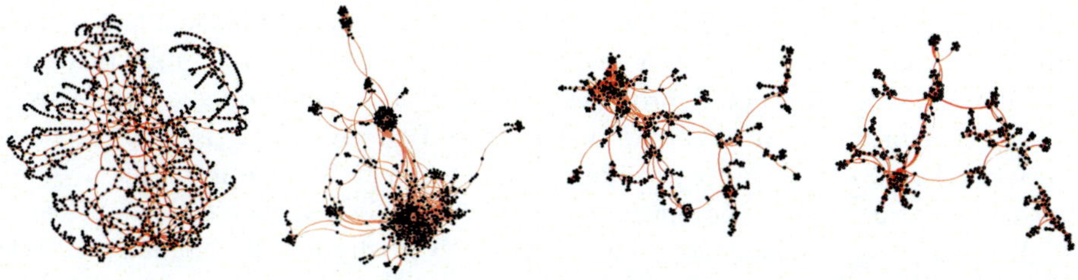

(a) Euroroad network (b) FB-Pages (c) Bio-diseasome (d) CA-netscience

Fig. 4. Visualization in Gephi of four real networks topologies, where a darker color indicates a larger node degree.

Table 2. GrMFPT for six real networks with various random walks, where (T) indicates results with theoretical expressions and (N) with numerical simulations.

	U-RW	ID-RW	2H-RWM	ID-RWM	F-RWM	P-RWM	PID-RWM
Internet (N)	19385	178916	18293	23410	17318	16443	20260
Wikipedia (N)	22974882	11342	1269384	10930	29894802	9086546	11857
Euroroad (N)	9246	12854	5489	2968	2742	2714	2900
Euroroad (T)	9243	12762	5485	2954	2760	2716	2922
FB-Pages (N)	3516	3879	1673	1588	2422	1845	1450
FB-Pages (T)	3521	3855	1676	1568	2401	1846	1453
CA-netscience (N)	1895	4747	1287	2326	1409	1046	2488
CA-netscience (T)	1891	4742	1297	2354	1409	1062	2694
Bio-diseasome (N)	3526	8536	2114	4277	2471	1663	3246
Bio-diseasome (T)	3488	8503	2119	4322	2474	1662	3300

For example, P-RWM is better than PID-RWM and ID-RWM, which was often not the case in the generated networks. P-RWM is better for most networks, but for Wikipedia it is significantly worse than ID-RWM and PID-RWM and it also fails behind them for FB-pages. Another interesting observation is that from all considered real networks only Wikipedia is directed and has a very high average node degree, hence, the inverse degree biased group of random walks show best results. In this network the forward going behavior is naturally enforced, while the inverse degree biasing flattens the visiting probabilities and speeds up the search.

5 Conclusion

We studied various types of graph searching algorithms based on biased random walks using local information and a one-hop memory, which can be applied in modelling real-world phenomena and solving various problems. The results calculated with the given theoretical expressions match those with the numerical simulations both for generated and real networks. Generally biasing can be helpful, particularly in undirected networks, however, it should be applied carefully as different strategies could produce varying results depending on the specific network properties. For example, biasing based on inverse degree can be useful in networks with a scale-free property, but it can be unfavourable in networks with large transitivity. Moreover, the application in real networks can lead to slightly different results from what is obtained in supposedly similar generated networks. As a future work one can expand the study and include multiple random walkers, however, in this way the transition and state matrices would increase exponentially with the number of walkers. Another possible direction is considering memory in random walk with restart or teleportation, or random walk on hypergraphs.

Acknowledgement. This research was partially supported by the Faculty of Computer Science and Engineering, at the Ss. Cyril and Methodius University in Skopje, N. Macedonia.

References

1. Austerweil, J., Abbott, J.T., Griffiths, T.: Human memory search as a random walk in a semantic network. In: Advances in Neural Information Processing Systems, vol. 25. Curran Associates, Inc. (2012)
2. Backstrom, L., Leskovec, J.: Supervised random walks: predicting and recommending links in social networks. In: Proceedings of the 4th ACM International Conference on Web Search and Data Mining, pp. 635–644 (2011)
3. Basnarkov, L., Mirchev, M., Kocarev, L.: Persistent random search on complex networks. In: Trajanov, D., Bakeva, V. (eds.) ICT Innovations 2017. CCIS, vol. 778, pp. 102–111. Springer, Cham (2017)
4. Basnarkov, L., Mirchev, M., Kocarev, L.: Random walk with memory on complex networks. Phys. Rev. E **102**(4), 042315 (2020)
5. Berenbrink, P., Cooper, C., Elsässer, R., Radzik, T., Sauerwald, T.: Speeding up random walks with neighborhood exploration. In: Proceedings of the 21st Annual ACM-SIAM Symposium on Discrete Algorithms, pp. 1422–1435. SIAM (2010)
6. Bonaventura, M., Nicosia, V., Latora, V.: Characteristic times of biased random walks on complex networks. Phys. Rev. E **89**(1), 012803 (2014)
7. Cao, X., Wang, Y., Li, C., Weng, T., Yang, H., Gu, C.: One-step memory random walk on complex networks: an efficient local navigation strategy. Fluct. Noise Lett. **20**(05), 2150040 (2021)
8. Carletti, T., Battiston, F., Cencetti, G., Fanelli, D.: Random walks on hypergraphs. Phys. Rev. E **101**(2), 022308 (2020)
9. Fronczak, A., Fronczak, P.: Biased random walks in complex networks: the role of local navigation rules. Phys. Rev. E **80**(1), 016107 (2009)
10. Glonek, M., Tuke, J., Mitchell, L., Bean, N.: Semi-supervised graph labelling reveals increasing partisanship in the united states congress. Appl. Netw. Sci. **4**(1), 1–18 (2019)
11. Goh, K.I., Cusick, M.E., Valle, D., Childs, B., Vidal, M., Barabási, A.L.: The human disease network. Proc. Natl. Acad. Sci. **104**(21), 8685–8690 (2007)

12. Grinstead, C.M., Snell, J.L.: Introduction to probability. American Mathematical Society (2012)
13. Grover, A., Leskovec, J.: Node2vec: scalable feature learning for networks. In: Proceedings of the 22nd ACM SIGKDD International Conference on Knowledge Discovery and Data Mining, pp. 855-864. ACM (2016)
14. Leskovec, J., Kleinberg, J., Faloutsos, C.: Graphs over time: densification laws, shrinking diameters and possible explanations. In: Proceedings of the 11th ACM SIGKDD International Conference on Knowledge Discovery and Data Mining, pp. 177–187. ACM (2005)
15. Leskovec, J., Sosič, R.: Snap: a general-purpose network analysis and graph-mining library. ACM Trans. Intell. Syst. Technol. **8**(1), 1–20 (2016)
16. Newman, M.E.: Finding community structure in networks using the eigenvectors of matrices. Phys. Rev. E **74**(3), 036104 (2006)
17. Nikolentzos, G., Vazirgiannis, M.: Random walk graph neural networks. Adv. Neural Inf. Process. Syst. **33**, 16211–16222 (2020)
18. Noh, J.D., Rieger, H.: Random walks on complex networks. Phys. Rev. Lett. **92**(11), 118701 (2004)
19. Page, L., Brin, S., Motwani, R., Winograd, T.: The pagerank citation ranking: Bringing order to the web. Technical report, Stanford InfoLab (1999)
20. Patel, R., Carron, A., Bullo, F.: The hitting time of multiple random walks. SIAM J. Matrix Anal. Appl. **37**(3), 933–954 (2016)
21. Riascos, A.P., Boyer, D., Herringer, P., Mateos, J.L.: Random walks on networks with stochastic resetting. Phys. Rev. E **101**(6), 062147 (2020)
22. Rossi, R.A., Ahmed, N.K.: The network data repository with interactive graph analytics and visualization. In: Proceedings of the 29th AAAI Conference on Artificial Intelligence (2015). https://networkrepository.com/
23. Rozemberczki, B., Davies, R., Sarkar, R., Sutton, C.: Gemsec: graph embedding with self clustering. In: Proceedings of the 2019 IEEE/ACM International Conference on Advances in Social Networks Analysis and Mining, pp. 65–72. ACM (2019)
24. Seneta, E.: Non-negative Matrices and Markov Chains. Springer, Heidelberg (2006)
25. Šubelj, L., Bajec, M.: Robust network community detection using balanced propagation. Eur. Phys. J. B **81**(3), 353–362 (2011)
26. Tejedor, V., Bénichou, O., Voituriez, R.: Global mean first-passage times of random walks on complex networks. Phys. Rev. E **80**(6), 065104 (2009)
27. West, R., Leskovec, J.: Human wayfinding in information networks. In: Proceedings of the 21st International Conference on World Wide Web, pp. 619–628 (2012)
28. West, R., Pineau, J., Precup, D.: Wikispeedia: an online game for inferring semantic distances between concepts. In: 21st International Joint Conference on Artificial Intelligence (2009)
29. Zhu, X., Ghahramani, Z., Lafferty, J.D.: Semi-supervised learning using gaussian fields and harmonic functions. In: Proceedings of the 20th International Conference on Machine Learning (ICML-2003), pp. 912–919 (2003)

Network Entropy as a Measure of Socioeconomic Segregation in Residential and Employment Landscapes

Nandini Iyer[1(✉)], Ronaldo Menezes[1,2], and Hugo Barbosa[1]

[1] BioComplex Laboratory, Computer Science, University of Exeter, Exeter, UK
niyer@biocomplex.org, {r.menezes,h.barbosa}@exeter.ac.uk
[2] Computer Science, Federal University of Ceará, Fortaleza, Brazil

Abstract. Cities create potential for individuals from different backgrounds to interact with one another. It is often the case, however, that urban infrastructure obfuscates this potential, creating dense pockets of affluence and poverty throughout a region. The spatial distribution of job opportunities, and how it intersects with the residential landscape, is one of many such obstacles. In this paper, we apply global and local measures of entropy to the commuting networks of 25 US cities to capture structural diversity in residential and work patterns. We identify significant relationships between the heterogeneity of commuting origins and destinations with levels of employment and residential segregation, respectively. Finally, by comparing the local entropy values of low and high-income networks, we highlight how disparities in entropy are indicative of both employment segregation and residential inhomogeneities. Ultimately, this work motivates the application of network entropy to understand segregation not just from a residential perspective, but an experiential one as well.

Keywords: Commuting networks · Network entropy · Socioeconomic inequality

1 Introduction

Often, the prosperity of urban areas is understood as a function of economic input and output [12]. However, experiential variables such as social inclusion have been shown to fuel the productivity of cities [5,15]. Accordingly, investing in improving the diversity of cities can lead to positive socioeconomic outcomes, fostering innovation and entrepreneurship [13]. The urban experience, in terms of mobility, is largely comprised of trips to work [8], underscoring the importance of considering social inclusion not only from the residential dimension, but the employment perspective as well. In this sense, mobility, specifically commuting patterns, can be viewed as a potential way to improve diversity and integration by creating points of social inclusion in employment areas [7].

Networks are particularly useful for analysing commuting behaviours, as they capture structural patterns that other approaches may overlook [10]. Specifically,

A. S. Teixeira et al. (Eds.): CompleNet 2023, SPCOM, pp. 26–37, 2023.
https://doi.org/10.1007/978-3-031-28276-8_3

network entropy can capture the concentration of labour supply and demand as well as the level of diversity of where workers are commuting to or from [11]. Entropy has been used in commuting networks to explain economic growth [6], identify spatial inequalities [9], and measure social assortativity [1]. However, the majority of previous research on this topic analyses the commuting networks of an entire population. Thus, we consider not only the commuting networks of an entire population, but also commuting networks comprised of individuals from particular socioeconomic groups. Disaggregating commuting networks by demographics allows us to study whether disparities in structural diversity could serve as an indicator of social exclusion.

In this paper, we explore the intersection of socioeconomic segregation in cities and structural diversity in commuting patterns by applying an information-theoretic approach to mobility networks. Specifically, we leverage measures of global and local entropy of commuting networks to clarify whether demographic disparities in reliance on commuting origins and destinations correspond with levels of segregation on a residential and employment level. Our analyses of the commuting networks of 25 US cities indicate that, for high-income workers, higher levels of residential concentration correlate with higher employment segregation. Employment segregation strays from residential segregation, in that it measures the level of segregation based on the individuals that work in a region, rather than individuals that live there. Low-income workers, on the other hand, tend to commute to a larger range of workplace areas. Finally, our network approach to segregation reveals facets of structural socioeconomic inhomogeneities in commuting patterns beyond what traditional measures of segregation can capture.

2 Methods and Data

2.1 Data

We use the LEHD Origin-Destination Employment Statistics (LODES) dataset from the United States Census's 2019 Longitudinal Employer-Household Dynamics (LEHD) program [4]. The dataset captures the residential patterns of the surveyed workforce by measuring the number of individuals commuting from one census block group to another. We evaluate residential-employment trends on a census tract level. Census tracts are statistical partitions of counties that contain anywhere from 1,200 to 8,000 residents. We use the LODES data to construct commuting networks for 25 cities in the United States, where every node reflects a census tract and directed, weighted edges depict the number of individuals commuting from one tract to another. The 25 cities we selected cover a wide range of population sizes and socioeconomic characteristics. The LODES dataset also provides information about how the total commutes from a pair of census tracts are distributed across lower, middle, and higher-income demographics. The low-income group consists of individuals earning less than \$1,250 per month, while the minimum monthly income for the high-income group is \$3,333. In addition to building the entire commuting network of a city, we also build a low, middle, and high-income network, which have the same nodes as the network for the entire city. However, the networks for each socioeconomic group, which we refer to as disaggregated networks, have different edge weights depending on the number of people in a socioeconomic group that commute between a pair of tracts.

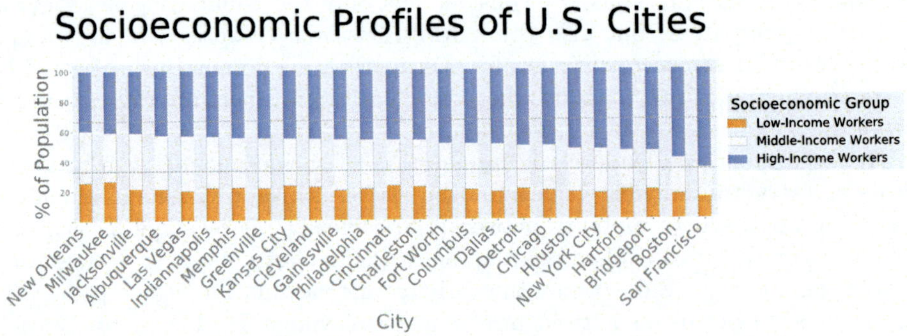

Fig. 1. Workforce distribution by income level split into three socioeconomic groups (low, middle and high-income) across 25 US cities.

Figure 1 captures the socioeconomic makeup of each city, according to the LODES data. The dashed lines indicate what an even distribution across demographics would look like. Thus, we observe that, within the context of the LODES dataset, cities such as Boston and San Francisco have skewed representation of socioeconomic groups. To measure levels of residential segregation, we use Table B19001 from the 2019 American Community Survey 5-Year Estimates, which measures the population distribution across income brackets for each census tract [3].

2.2 Network Entropy in Commuting Networks

Throughout our analyses, we use the term *commuting destinations* to refer to the workplaces that a residential population commutes to, while *commuting origins* describe the residential areas from which an employment area's workforce commutes. Furthermore, *labour supply* refers to employment areas that supply jobs, while *labour demand* reflects the employable population of a region. We use Shannon's entropy, which captures the level of information that can be extracted given a probability distribution, as a measure of diversity in commuting destinations and origins [14]. Moreover, network entropy can be applied at different network resolutions and can focus on the commuting in-flow to a work area or the commuting out-flow from a residential region. Global entropy of in-flow and out-flow captures the urban concentration of labour supply and demand, respectively, by characterising the degree of monocentricity. That is, when a city has one area that supplies most of the labour opportunities, there is less uncertainty in predicting commuting destinations, corresponding with a lower global in-flow entropy value for the entire city. Thus, global in-flow entropy (H_{GN}^{in}) leverages the node strength, $\sum_i p_{ij}$, of all incoming commutes to each tract, j in a city:

$$H_{GN}^{in} = \frac{-\sum_{\forall j} (\sum_{\forall i} p_{ij}) \, log \, (\sum_{\forall i} p_{ij})}{log(n)}, \tag{1}$$

where n is the total number of tracts in a city. Global out-flow entropy captures the distribution of labour demand, such that low entropy values depict a scenario in which most workers live in a few areas and high entropy values indicate where residential locations of workers are more evenly distributed across nodes. In contrast to H_{GN}^{in}, global out-flow entropy (H_{GN}^{out}) calculates the out-degree node strength, $\sum_j p_{ij}$, for all census tracts, i in a city.

Global entropy defines a city with respect to how labour supply or demand is distributed across all census tracts. Meanwhile, local entropy defines each tract based on how evenly distributed all its incoming or outgoing commutes are. A high local in-entropy for a census tract implies that the commuting origins for individuals who work in that tract are evenly distributed across all the potential origins. Local in-flow entropy (H_L^{in}) for a census tract j accounts for the probability, $p_{(i|j)}$, of tract j receiving commutes from a census tract i, for all possible commuting origins, i:

$$H_L^{in} = \frac{-\sum_{\forall i} \frac{p_{ij}}{p_j} log \frac{p_{ij}}{p_j}}{log(n-1)}. \tag{2}$$

Local out-flow entropy (H_L^{out}) can be defined similarly, except rather than consider the probability $p_{(i|j)}$ for incoming commutes, it calculates the probability, $p_{(j|i)}$ of tract i sending commuters to a census tract j, for all possible workplaces j. In this manner, we leverage network entropy measures on a global and local scale to understand how commuting characteristics of low versus high income workers indicate levels of segregation not only in terms of where each demographic group lives, but also in the context of the areas where they work. The denominators in Eqs. 1 and 2 serve to normalise the entropy values for cities of varying sizes, based on the number of tracts in a city (Eq. 1), and the focal node's maximal possible degree (Eq. 2).

2.3 Characterising Segregation in Urban Landscapes

We analyse features of the urban space to better understand the mechanisms that drive the identified differences in commuting networks across demographics. For such, we characterise urban areas in terms of the concentration of socioeconomic groups, using the Index of Concentration at the Extremes (ICE), proposed by Douglas Massey in an attempt to describe a region using both spatial attributes and the characteristics of polarised demographics [2]. While most segregation metrics account for how segregated the minority group is in relation to the entire population, ICE incorporates both ends of the demographic spectrum:

$$\text{ICE}_i = \frac{A_i - P_i}{T_i}, \tag{3}$$

where the ICE for a census tract i is defined as the difference between the number of affluent residents, A_i, and the residents under the poverty line, P_i, over the entire population, T_i. While the numerator captures the imbalance between the extremes, the denominator expresses the degree of imbalance in relation to the entire population of tract i. Thus, ICE aims to measure the imbalance between affluence and poverty by measuring the concentration of both the extremely disadvantaged and advantaged in a given population. In a similar vein, we can use Eq. 3 to measure the level of segregation for a census tract, i, from an employment perspective, defining A_i and P_i as the number of workers commuting to tract i from the high and low-income group, respectively. Thus, T_i captures the total number of individuals commuting to i, regardless of demographics. Accordingly, we can use ICE to capture segregation levels at both a residential (ICE_{res}) and employment (ICE_{emp}) scale. We use the LODES dataset to calculate ICE_{emp} as it provides unique information regarding demographic characteristics of a workforce. Meanwhile, We use the ACS median household income distributions to calculate residential segregation as it provides a high granularity of income-levels to characterise tracts by its residents.

3 Results

3.1 Global Entropy and City-Level Analyses

We begin by applying global entropy measures to the commuting networks of the 25 different cities. We reiterate that because entropy values are normalised with respect to network size, we can make comparisons not only between cities, but also between commuting networks of different socioeconomic groups. Figure 2 elucidates how, for every city, the global entropy of commuting out-flows is consistently larger than the global entropy for commuting in-flows. This pattern is to be expected, as residential locations are known to be more evenly distributed than employment hubs [11]. Notably, the lower values of global in-flow entropy in cities such as New Orleans, San Francisco, and Boston indicate the presence of larger employment hubs.

By disaggregating the commuting networks based on workers' economic profiles, we can analyse the networks of low-income and high-income workers separately. Interestingly, within the cities we analyse, the global in-flow entropy of high-income commuters is persistently lower than that of the low-income group. These lower values imply less structural diversity in high-income commuting origins, which translates to a higher level of monocentricity for high-income jobs. What is clear is that there is a distinct difference in global entropy values for commuting in-flows when considering networks of different socioeconomic groups. Whether these differences are indicative of socioeconomic inequality is explored in the following sections.

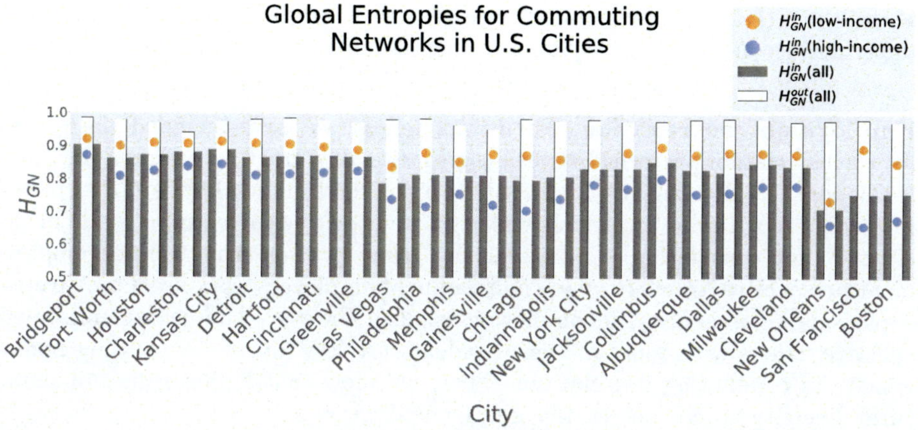

Fig. 2. Global entropy values for 25 US cites. The grey and white bar capture global entropy of the entire commuting network for in-flow and out-flow commutes, respectively. The orange point measures H^{in}_{GN} for the low-income network, while the purple point does the same for the high income network.

3.2 Local Entropy as a Measure of Segregation

In this section, we aim to disentangle whether overall structural diversity entails other forms of diversity, such as lower levels of segregation. In doing so, our goal is to clarify if the monocentricity of job opportunities, which is more present in high-income networks, is a privilege or a burden.

Residential Segregation. To better understand the differences in how housing and employment landscapes intersect for socioeconomic groups, we evaluate whether a relationship exists between residential segregation and the diversity of employment destinations for residents of a particular tract. The fourth column of Table 1 lists the Pearson correlation coefficients when comparing the local out-flow entropy (H_L^{out}) with the residential ICE value (Eq. 3) of census tracts in a city. We recall that the higher the ICE_{res} value, the larger the proportion of high-income residents in that area. All but three of the 25 cities have a significant, negative correlation. This reveals that census tracts characterised by more affluent residents tend to have lower local diversity values (i.e., concentrate commutes to fewer tracts), which indicate their dependence on particular tracts for supplying labour to their residents. On one hand, one can argue that these higher diversity values for less affluent tracts makes them less vulnerable to any shortages in labour supply, such that if an employment location stops providing opportunities, they have other options of commuting destinations. However, this negative correlation could also indicate the presence of inequalities in the housing landscape. Thus, we proceed to analyse the opposite dynamic, comparing the diversity of commuting origins (H_L^{in}) to the degree of employment segregation (ICE_{emp}), to explore whether the identified negative correlations imply socioeconomic disparity on a residential or employment level.

Employment Segregation. For completeness, we have included the results for the correlation patterns between segregation of a tract's workforce (ICE_{emp}, Eq. 3) and the diversity of commuting origins for that workforce (H_L^{in}) in the fifth column of Table 1, which shows significant positive correlations for 20 of the cities. When considering the high-income group, we see that, from an employment perspective, tracts in which more affluent employees work are more diverse, expressing heterogeneity in commuting origins. However, from a residential perspective, tracts with more affluent residents are less diverse, expressing homogeneity in commuting destinations. Thus, we show that higher values of structural diversity do not necessarily imply an advantage.

This section elucidated how the structural diversity of census tracts often corresponds with their urban characteristics. We measure the local in and out-flow entropy of entire commuting networks to highlight how diversity of the employment landscape can be reflective of employment and residential segregation. The next section wraps up this analysis by disaggregating the entire commuting network of each city into low and high-income networks. This allows us to examine how unequal labour distribution may be exacerbating existing inequalities.

3.3 Socioeconomic Disparities in Diversity of Commuting Origins

In this section, we illustrate how local in-flow entropy measures of disaggregated networks can capture experienced segregation. For improved readability and due to space constraints, we show the results for four representative cities of different sizes and socioeconomic profiles (Milwaukee, San Francisco, New York City and Detroit). Nevertheless, our findings are based on the analyses of the 25 cities.

We proceed, analysing how differences in heterogeneity of commuting origins coincide with levels of employment segregation. We begin by splitting the commuting network of the city into separate networks that measure the residential-work patterns of socioeconomic groups separately. In doing so, we can understand the extent to which the diversity of commuting origins differs between the low-income and high-income workers in a census tract. Panels A-D in Fig. 3 plot the local in-flow entropies of every tract in the low-income network ($H_{L,lo}^{in}$) against their respective entropy values in the high-income network ($H_{L,hi}^{in}$), for four of the analysed cities. The black line expresses the case in which a census tract has equal diversity of commuting origins in both socioeconomic networks, which we see a few cases of in New York City and Detroit, indicated by the white points on the diagonal. Orange points reflect census tracts in which the low-income individuals have a more even distribution of commuting origins than the high-income workforce. The purple points capture census tracts with the opposite characteristics: greater diversity in residential locations for the affluent workforce. We observe that most census tracts in San Francisco and New York

Table 1. General network properties and Pearson correlation coefficients for different forms of local entropy and socioeconomic segregation. The last column refers to local entropy differences between the disaggregated networks, discussed in Sect. 3.3. Boldface is used to indicate significant correlations, with asterisks reflecting p-value.

City	Network properties		H_L^{out} vs. ICE_{res}	H_L^{in} vs. ICE_{emp}	ΔH_L^{in} vs. ICE_{emp}
	Nodes	*Edges*	r^a	r^a	r^a
Charleston	85	6,162	**–0.313****	0.116	**0.453*****
San Francisco	196	25,886	**–0.373*****	**0.614*****	**0.341*****
Gainesville	56	2,803	–0.243	**0.313***	**0.496*****
Greenville	111	10,431	**–0.386*****	0.033	**0.596*****
Albuquerque	153	18,871	**–0.385*****	0.116	**0.632*****
New Orleans	176	13,680	**–0.309*****	0.140	**0.843*****
Houston	921	398,876	**–0.287*****	–0.018	**0.716*****
Boston	204	20,608	**–0.401*****	**0.652*****	**0.646*****
Indianapolis	224	33,147	**–0.494*****	**0.231*****	**0.754*****
Las Vegas	487	124,083	0.027	**0.093***	**0.818*****
Philadelphia	384	69,364	**–0.547*****	**0.289*****	**0.754*****
Columbus	347	76,995	**–0.486*****	**0.272*****	**0.735*****
Hartford	224	33,107	**–0.507*****	**0.430*****	**0.710*****
Jacksonville	173	23,610	**–0.441*****	**0.473*****	**0.722*****
Cincinnati	222	32,801	**–0.185****	**0.306*****	**0.751*****
Milwaukee	297	50,210	**–0.517*****	**0.377*****	**0.829*****
Cleveland	446	85,609	**–0.182*****	**0.330*****	**0.834*****
Bridgeport	210	28,259	**–0.425*****	**0.278*****	**0.665*****
Fort Worth	357	76,883	**–0.483*****	**0.280*****	**0.798*****
Memphis	221	31,507	**–0.135***	**0.212****	**0.816*****
Chicago	1,318	441,406	**–0.287*****	**0.358*****	**0.855*****
New York City	2,164	990,302	**–0.478*****	**0.455*****	**0.837*****
Detroit	1,163	385,185	**0.139*****	**0.527*****	**0.865*****
Dallas	529	129,486	**–0.569*****	**0.350*****	**0.830*****
Kansas City	283	49,377	**–0.165****	**0.328*****	**0.781*****

[a]*p<0.05; **p<0.01; ***p<0.001

City tend to have higher homogeneity in commuting origins for low-income workers than compared to high-income workers. This comparison opens the gateway into using local entropy values to extend our understanding of segregation from a residential dimension to an employment one as well.

In order to accomplish this, we evaluate how socioeconomic disparities in the homogeneity of residential locations for a region relate to the demographic balance of that region's workforce. For each tract in a city, we compare a census

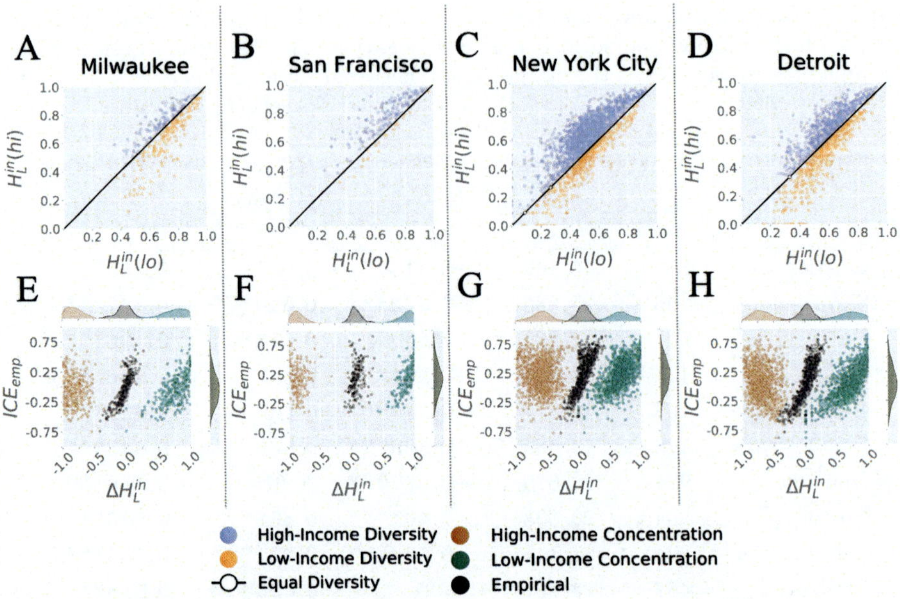

Fig. 3. Local in-flow entropies for high and low-income networks in 4 US cities. Panels A-D compare entropy values for tracts in the low and high-income network, with points below the diagonal reflecting tracts in which low income workers have more diverse commuting origins. Panels E-H show how the differences in these values (black points) compare to null models derived from Fig. 4

tract's in-flow entropy value in the high and low-income networks:

$$\Delta H_L^{in}(i) = H_{L,hi}^{in}(i) - H_{L,lo}^{in}(i), \tag{4}$$

where $H_{L,hi}^{in}(i)$ captures the local in-flow entropy of census tract i in the high-income commuting network, whereas $H_{L,lo}^{in}(i)$ describes the in-flow entropy for i in the low-income commuting network. Thus, $\Delta H_L^{in}(i)$ can range from -1 to 1, where negative values represent more heterogeneity of commuting origins for the low-income population. Positive values capture scenarios in which higher-income workers have more heterogeneous commuting origins than lower-income workers.

Comparing Local Entropies in Disaggregated Networks. For each of the 25 cities, we find significant positive correlations between the difference in local entropy values ($\Delta H_L^{in}(i)$, Eq. 4) to the level of segregation in employment areas (ICE_{emp}, Eq. 3). These correlations are outlined in the last column of Table 1. If we consider the local in-flow entropy of tract i in the high-income network, we can measure how evenly the total number of affluent individuals commuting to i is distributed across all other census tracts in the city. We use Fig. 4 to explain how entropy values are not strictly correlated with in-degrees.

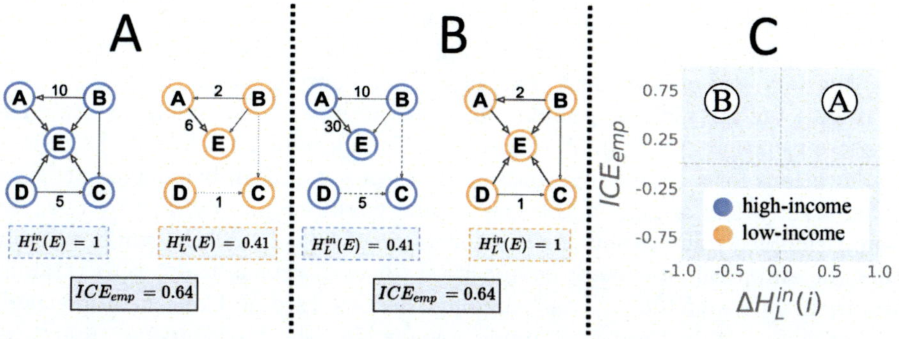

Fig. 4. Toy example highlighting the distinction between $\Delta H_L^{in}(i)$ and ICE_{emp}. Panel A shows concentration of high-incoming commuting origins, while B captures homogeneous origins for the low-income group. Panel C conveys how the two scenarios of commuting diversity relate to employment segregation.

The purple network captures the high-income commuting network and the orange network reflects that of the low-income workers. We set the total number of individuals working in node E to be 50. For Scenarios A and B, we can observe that the socioeconomic composition of individuals working in node E remains the same (40 high-income workers, 8 low-income workers). Thus, values of employment segregation (ICE_{emp}) are consistent throughout the examples. What changes across the scenarios are the values of H_L^{in}, which describe how evenly commuting origins are distributed across other nodes in the network. Scenario A depicts a case where low-income commutes are more concentrated and high-income commuting origins are more diverse. Meanwhile, Scenario B captures homogeneity in high-income commuting origins and heterogeneous commuting origins for the low-income workers in the node E. We can, then, understand that $\Delta H_L^{in}(i)$ serves to measure differences in heterogeneity of commuting origins between the high and low income networks. The right most panel in Fig. 4 uses Scenarios A and B to elucidate how positive correlations between ICE_{emp} and $\Delta H_L^{in}(i)$ are non-trivial.

Null Model Comparisons. The black scatter plots in Fig. 3E-H illustrate the aforementioned positive correlations between socioeconomic differences in the heterogeneity of residences ($\Delta H_L^{in}(i)$) and employment segregation ICE_{emp}, for four cities. We use null models, inspired from the examples in Scenario A and B of Fig. 4, to emphasise how the observed positive correlations are not a result of measuring related network attributes. The null models elucidate how the correlations between employment segregation and local entropy of networks are not always significant or positive. Both null models retain the in-degree of the disaggregated, empirical commuting networks, only changing how evenly a node's incoming edges are distributed across other nodes in the network.

The Low-Income Concentration null model, represented by the blue scatter plot, hypothesises that the observed positive correlation between commuting origin diversity and employment segregation is a consequence of greater diversity of origins for the affluent population than the low-income group, captured by positive values of $\Delta H_L^{in}(i)$. We construct this null model, such that the residential locations for a tract's high-income workforce is uniformly sampled across all other tracts. The origins of the low-income workforce are defined by one, randomly sampled tract, producing smaller values of $H_{L,lo}^{in}(i)$ All the while, we maintain the empirical workplace composition for each census tract, thus retaining empirical values of ICE_{emp} and aligning the Low-Income Concentration model with Scenario A in the toy example. Meanwhile, the High-Income Concentration model, inspired by Scenario B in the toy example, and the brown scatter plot in Fig. 3, capture negative values of $\Delta H_L^{in}(i)$ by simulating high diversity of low-income commuting origins and homogeneity for high-income origins.

We emphasise that, in both null models, the only difference from the empirical commuting network is from the distribution of incoming edges to a tract. The equivalent distributions for ICE_{emp}, shown along the y-axis, demonstrate how measures of employment segregation remain consistent across the empirical and null model scenarios, despite having disparities in residential-work patterns. On the other hand, the distributions along the x-axis, illustrate how $\Delta H_L^{in}(i)$ can capture these structural inequalities by identifying tracts in which socioeconomic groups express stark differences in their dependence on commuting origins. While the empirical scatter plot does not show signs of extreme structural disparities, as expressed by the null models' scatter plots, the positive correlations indicate that areas which have more heterogeneity of residential locations for a particular socioeconomic group tend to be areas in which that socioeconomic group is more concentrated. The sign of $\Delta H_L^{in}(i)$ specifies which demographic is segregated, with values less than zero implying low-income employment segregation. Thus, this section highlights how measuring disparities in structural diversity of disaggregated networks can expose dimensions of segregation in residential-workplace dynamics, that conventional metrics of segregation may overlook.

4 Conclusion

This work highlights how network entropy can be used to capture sociodemographic inequalities in commuting patterns that classical segregation metrics fail to detect. We compare global in-flow entropies of 25 commuting networks across the U.S. to identify cities, such as San Francisco and Boston, that have higher degrees of monocentricity.

Then, by incorporating local entropy measures for the entire commuting network, we uncover that census tracts with a higher concentration of affluence have residents that travel to more homogenous workplaces. Meanwhile, tracts that attract a higher-income workforce express trends of heterogeneity in commuting origins. Finally, by splitting cities' commuting networks into high and low-income networks, we demonstrate the strength of network entropy in identifying disparities in resident-workplace trends across socioeconomic groups. We find that larger differences in node-level entropies, which measures socioeconomic disparities of a work force in terms of their residential distribution, correspond with higher levels of employment segregation.

This brings into question what has long been deliberated in sociological fields: how the consequences of segregation may change depending on which demographic is segregated. Future work can examine how these disparities in commuting flows reflect in other aspects of urban life. Specifically, other null models can be explored to understand what mechanisms may be fuelling the strong correlation between entropy and segregation. Moreover, this framework can be applied to other mobility networks in the context of various demographic dimensions, such as gender or age.

References

1. Bokányi, E., Juhász, S., Karsai, M., Lengyel, B.: Universal patterns of long-distance commuting and social assortativity in cities. Sci. Rep. **11**(1), 1–10 (2021)
2. Booth, A., Crouter, A.C.: The prodigal paradigm returns: ecology comes back to sociology. In: Does it take a village?, pp. 53–60. Psychology Press (2001)
3. Bureau, U.C.: 2019 American community survey 5-year estimates, table b19001 (2022). Accessed 15 Dec 2022
4. Bureau, U.C.: LEHD origin-destination employment statistics data (2002-2019) (2022). Accessed 15 Dec 2022. Longitudinal-Employer Household Dynamics Program, LODES 7.5
5. Diaz, R., Garrido, N., Vargas, M.: Segregation of high-skilled workers and the productivity of cities. Regional Sci. Policy Pract. **13**(5), 1460–1478 (2021)
6. Goetz, S.J., Han, Y., Findeis, J.L., Brasier, K.J.: Us commuting networks and economic growth: measurement and implications for spatial policy. Growth Chang. **41**(2), 276–302 (2010)
7. Hackl, A.: Mobility equity in a globalized world: reducing inequalities in the sustainable development agenda. World Dev. **112**, 150–162 (2018)
8. Jiang, S., Yang, Y., Gupta, S., Veneziano, D., Athavale, S., González, M.C.: The TimeGeo modeling framework for urban mobility without travel surveys. Proc. Natl. Acad. Sci. **113**(37), E5370–E5378 (2016)
9. Lenormand, M., Samaniego, H., Chaves, J.C., da Fonseca Vieira, V., da Silva, M.A.H.B., Evsukoff, A.G.: Entropy as a measure of attractiveness and socioeconomic complexity in Rio de Janeiro metropolitan area. Entropy **22**(3), 368 (2020)
10. Louail, T., et al.: Uncovering the spatial structure of mobility networks. Nat. Commun. **6**(1), 1–8 (2015)
11. Marin, V., Molinero, C., Arcaute, E.: Uncovering structural diversity in commuting networks: global and local entropy. Sci. Rep. **12**(1), 1–13 (2022)
12. OECD: Compendium of productivity indicators 2013 (2013)
13. Qian, H.: Diversity versus tolerance: the social drivers of innovation and entrepreneurship in us cities. Urban Stud. **50**(13), 2718–2735 (2013)
14. Shannon, C.E.: A mathematical theory of communication. Bell Syst. Techn. J. **27**(3), 379–423 (1948)
15. Veneri, P., Comandon, A., Garcia-López, M.À., Daams, M.N.: What do divided cities have in common? an international comparison of income segregation. J. Reg. Sci. **61**(1), 162–188 (2021)

Community Structure in Transcriptional Regulatory Networks of Yeast Species

Fábio Cruz[1,2,3(✉)], Pedro T. Monteiro[1,2], and Andreia Sofia Teixeira[3]

[1] Instituto Superior Técnico, Universidade de Lisboa, Lisboa, Portugal
fabio.rocha.cruz@tecnico.ulisboa.pt
[2] INESC-ID, Lisboa, Portugal
[3] LASIGE, Departamento de Informática, Faculdade de Ciências, Universidade de Lisboa, Lisboa, Portugal

Abstract. The genomic expression of living organisms is controlled by complex transcriptional regulation. Transcriptional regulatory networks, composed by associations between transcription factors and target genes, are responsible for representing and controlling this gene expression and regulate the response of an organism to environmental changes. In this paper, we extend the study of these systems by applying different community detection algorithms on closely-related yeast transcriptional regulation networks to characterize their topological structure and understand if these methods are able to capture meaningful functional clusters of genes. We start by evaluating the accuracy and efficiency of a large group of algorithms by applying them to benchmark networks with ground-truth communities. We then apply the methods that had the best performance to the yeast networks to analyze the quality of the resulting structures, and then, assess the quality of the retrieved modules from a biological point of view using available annotated species' biological functions. Finally, we apply a multilayer community detection algorithm on multilayer networks, where each layer is an individual yeast network, and use available mappings between nodes of different species to successfully discover communities with genes that belong to different species but have similar biological functions. We conclude that the use of community detection algorithms to functionally characterize the modules of these networks might not be enough, suggesting the need of additional genetic information and possibly the use of alternative strategies to study these complex regulatory networks.

Keywords: Complex networks · Transcriptional regulatory networks · Community detection · Multilayer networks

1 Introduction

Transcriptional regulation is an important mechanism for controlling gene expression, which allows the organism to respond to changes in the environment by altering the production of specific gene products. This control is managed by transcription factors (TF) [12] and other proteins, that can activate or repress a wide variety of target genes (TG) to ensure they are expressed at

A. S. Teixeira et al. (Eds.): CompleNet 2023, SPCOM, pp. 38–49, 2023.
https://doi.org/10.1007/978-3-031-28276-8_4

the right time and in the right amount throughout the lifetime of a cell. These interactions between transcription factors and target genes define transcriptional regulatory networks, that contain information about the biological functionality of the organisms. These networks are of great biological importance [1,13], and the knowledge gathered from their structure and evolution can be used to analyze the regulatory networks of different organisms, thus facilitating the understanding of differential gene expression and bringing new developments to the scientific community. Despite this, there is still plenty of uncertainty regarding the structures and dynamics of this type of transcriptional regulatory networks. One of the most common approaches to study these networks is to analyze their structure through network science techniques. In network science, networks that result from complex real-life systems from many areas can be represented by graphs [2], such as citation networks, social networks, biological networks, among others. Transcriptional regulatory networks are no exception. Furthermore, it is possible to study the structure of those graphs to discover underlying functionality and dynamics that are not obvious at first sight. This can be done by detecting community structure through the use of algorithms for this purpose [6]. The study of communities has gained great importance in recent times. In the biological field, the identification of communities can be used to discover new structures associated with specific genetic functionalities previously unknown. Our main goal is to contribute to expand the knowledge of the community about this subject by characterizing the transcriptional regulatory networks of 10 closely-related species through a network science approach. To this end, we will study the community structure of these networks and relate it with the biological functionalities of the entities that compose them, and also relate the obtained information between the closely-related species. The transcriptional regulatory networks of these species are provided by YEASTRACT+ [15].

2 Preprocessing

Initially, we will introduce the networks from yeast species that we will be working on, as well as briefly analyze some of their structural characteristics. Furthermore, instead of blindly relying on some algorithms to detect communities in our yeast networks for reasons that are not related to their actual performance, like the popularity of the algorithm or of its creator, we will evaluate the accuracy and efficiency of a large group of algorithms by applying them to benchmark networks with built-in community structure.

2.1 Yeast Networks

The yeast networks are directed, where each origin node represents a transcription factor (TF) that is associated with a target gene (TG). These regulatory associations are essentially supported by DNA binding evidence retrieved from experimental data available, and/or expression evidence. Most of the links of the last-mentioned group can still be divided into four types, that are represented

Table 1. Yeast network properties. The properties shown are the number of nodes of each network, number of edges, number of transcription factors, average degree, average clustering coefficient, average path length and diameter, in this order.

Network	Nodes	Edges	TF	TG	$\langle k \rangle$	CC	APL	D
S. cerevisiae (Sc)	6 886	195 498	220	6 886	56.78	0.47	0.059	4
C. albicans (Ca)	6 015	35 687	118	6 015	11.87	0.28	0.034	5
Y. lypolytica (Yl)	5 288	9 238	5	5 288	3.49	0.36	0.001	4
C. parapsilosis (Cp)	3 381	6 986	11	3 380	4.13	0.25	0.003	4
C. glabrata (Cg)	2 133	3 508	40	2 116	3.29	0.05	0.002	6
C. tropicalis (Ct)	665	698	16	663	2.10	0.01	0.041	5
K. pastoris (Kp)	561	581	4	559	2.07	0.01	0.002	5
K. lactis (Kl)	111	126	10	106	2.27	0.15	0.014	2
Z. bailii (Zb)	32	31	1	31	1.94	0.00	0.031	2
K. marxianus (Km)	4	3	1	3	1.50	0.00	0.250	2

by a sign annotation in the network. Depending on the effect the TF has on the TG, edges can be positive if the TF is an activator, negative if it is a repressor or dual/unknown when it has a dual effect or the directionality of the effect is unknown [16]. In Table 1, we can observe some basic properties of these yeast networks, organized by the number of nodes, that help us understand how they are composed. The first thing that pops up is that some of the networks are severely underdocumented, as we can see from the number of links of the bottom five networks, all below a thousand. We can also see that the number of TFs does not always increase with the number of nodes and edges, meaning some networks are smaller than others but might have more relevant biological information, since origin nodes are usually the center of big hubs inside the networks. As for the clustering coefficient, we can observe that only the four largest networks have well-clustered groups of nodes while the others are small and sparse. The average path length has very low values in general since they are directed and mostly composed of big hubs where the TF can reach the TGs with a distance of one, but the TGs can not connect to each other, giving a distance of zero. Since the number of TGs in all networks is much larger than the number of TFs, the average path length will always tend to zero.

2.2 Benchmark Testing

Lancichinetti *et al.* proposed the LFR benchmark [10], which generates networks that follow a power-law distribution for the node degrees and community size to solve limitations found in previously introduced benchmarks, so we decided to initially test the algorithms with this benchmark. It also allows controlling several parameters for the generated graph, such as the mixing coefficient μ, which represents the desired average proportion of links between a node and nodes located outside its community, thus reflecting the amount of noise in the network. Regarding the set of community detection algorithms tested on the networks with ground-truth communities, we selected a mixture of popular and traditional methods and recently proposed ones, based on different strategies such as the optimization of modularity, random walks, and propagation of information. This set is composed of algorithms such as the Girvan-Newman [6,17], Louvain [3], Leiden [23], Label Propagation [19], and Infomap [22], amongst others. In order to assess the resemblance between the communities retrieved by the chosen algorithms and the LFR ground-truth communities, we decided to use the Normalized Mutual Information metric (NMI) [4], which is one of the most popular metrics for this purpose. Additionally, to avoid relying solely on a single metric, we also compared the different partitions with the Adjusted Rand Index (ARI) [8] and F1 [21] metrics. All of the algorithms, metrics, and networks were implemented with the help of NetworkX [7] and the Community Detection Library CDlib [20], a recent Python package that allows to extract, compare and evaluate communities from complex networks.

An alternative to these descriptive algorithms for community detection are inferential methods, which have been shown to be more robust in the analysis of network communities [18]. We chose to simplify our approach and not experiment with inferential methods since they can have a very high time complexity and also due to the lack of variety of implementations of these methods in the Python packages that we decided to utilize. Some of the methods being tested are not deterministic, which means their results can depend on specific random seeds or initial conditions used for their execution. In order to address this, we implemented a version of the consensus clustering strategy proposed by Lancichinetti and Fortunato [9] to generate stable results out of a set of partitions delivered by stochastic methods. We applied it to our stochastic methods but only did one iteration of their algorithm since the authors claim that one iteration is sufficient to lead to stable results. They show from their results that running the algorithm 100 times on the original network before generating the consensus matrix gives good results for most of the tested algorithms. The threshold of the consensus clustering and the mandatory parameters of some algorithms we selected were the ones that gave the best accuracy scores against the benchmark ground truth.

As we have shown previously, some of the yeast networks have a significant size, which means the time complexity of the algorithms is an important factor when choosing which ones we will use. For this reason, the first thing we tested was the execution times for the set of community detection algorithms on the LFR benchmark graph with 250, 500, and 1000 nodes. We observed that the worst-performing algorithms regarding time complexity for these graphs are GA-Net, Girvan-Newman, and CONGO. After this, we varied the LFR mixing coefficient μ to understand how the different methods would behave against a network with more or less entropy, and used the metrics mentioned before to compare and evaluate the communities calculated by the algorithms and the ground truth of the LFR network. In general, as expected, the higher the μ value, the worse the score values for all the metrics. Regarding the algorithms based on the optimization of modularity, we highlight Louvain and Leiden, whose scores were almost always better than the other algorithms for each graph and metric. EdMot [14], an alternative method proposed in 2019 that uses higher-order motifs and that is based on the Louvain algorithm, performed as well as the majority of the most popular methods. As for the remaining ones, we highlight Infomap and Walktrap, two algorithms based on random walks that had very good results for all different-sized graphs.

Even though we took some useful information from the results obtained on the LFR synthetic network generator, before choosing the final set of algorithms to apply to our yeast networks, we still also wanted to test the different methods on a real, ideally directed, network with ground-truth communities that follows a power-law degree distribution. For this reason, we tested them on the email-Eu-core network (Email network), retrieved from the Stanford Large Network Dataset Collection, a directed network with 1 005 nodes and 25 571 edges that was generated using email data from a large European research institution where each edge represents an email that was sent from one person to another. We ran the same tests on this network and, once again, the best performing algorithms were Louvain and Leiden, both having the best score values in all calculated measures, closely followed by Infomap and Walktrap.

3 Functional Characterization

In this section, we will take the best performers of the previous tests and apply them to the yeast networks in order to analyze some of their internal properties, and utilize biological annotations of the network nodes to understand how the topological structure of these networks is related to their biological functionality.

3.1 Structural Evaluation

The final set of methods that were chosen to analyze the yeast networks given the results of the previous experiences was the Louvain, Leiden, Walktrap, and Infomap algorithms. The remaining were discarded in this phase either for being too slow, having poor results after being tested against the benchmark networks, or even for existing better-performing options that follow the same principles.

We will now evaluate the communities retrieved by the algorithms for the yeast networks from a structural standpoint. First, analyzing the number of communities detected can be interesting to see how the four clustering alternatives are able to identify distinct groups of genes closely related to each other from a topological standpoint. A higher number of communities can indicate that the functional characterization of the network is more complete than others. We observed that most of the solutions were able to find at least more than one cluster for each network and that Infomap and Walktrap were able to find a larger diversity of clusters than Louvain and Leiden. A possible reason for this is the fact that both these algorithms use random walks and the networks are mainly composed of big hubs of nodes where one TF points to many TG's, meaning the random walkers will often get trapped in these hubs, thus considering them as distinct communities. If this is true, then the number of communities should increase with the fraction of TFs when compared to the number of nodes and edges of the network, which we can also confirm by the fact that more communities were found in the *Ca* network than the larger *Sc* network, and also found a higher number of communities in the *Cg* network when compared to some of its larger neighbours. After this, we also calculated the Newman-Girvan modularity [17] for the communities found by the algorithms. Generally speaking, we can say that most networks had low modularity scores, and the algorithm that was able to find the most modular partitions of the networks regarding this score function was Louvain, which makes sense since the implementation of this algorithm depends directly on the optimization of modularity. We also tested the significance metric [11] since high values of modularity do not always translate to a good community structure. The values for this score function were higher for larger networks, and Infomap and Walktrap gave higher significance scores than Louvain and Leiden, in particular for the two largest networks.

3.2 Genetic Classification

Our main goal is to study how the topological structure of the yeast networks is connected to the biological functionality of their components. However, unlike with the LFR and Email networks, we do not have ground-truth communities for these biological networks to compare with the partitions generated by the community detection algorithms. To help us solve this issue, we used available biological annotations for the genes that are represented by the network nodes, that were also recently considered by Monteiro *et al.* [16]. These annotations are labels that contain important information regarding the genetic functionality of the genes of each species. For instance, in the *Sc* network we can find a large variety of genetic functionalities, including some related to metabolism, protein functions, multidrug resistance, and many more. Our approach was to implement an algorithm that manually created communities for each network, using these biological annotations. We labeled each network node with one genetic classification and created clusters of nodes with similar labels. The nodes that did not have a known label were also grouped in the same cluster. This strategy allowed us to compare and evaluate the partitions of the algorithms on the original network with the artificial communities to understand if the algorithms are able to find groups of nodes with similar biological roles. After analyzing the scores of the same metrics previously tested, we saw that the partitions had very low similarity values regarding all measures, the majority below 10%. The exception was with Infomap, producing scores that ranged between 20% and 35% for the detected communities on the most documented directed networks, meaning it was the algorithm best capable of splitting the network nodes according to their biological information. One of the possible reasons for the low scores of the previous tests with the genetic classifications of the nodes is that the networks that we have been testing so far might have some noise. The regulatory associations in these biological networks have been registered based on numerous experimental setups that may be classified into two major groups:

Table 2. NMI scores for each algorithm against the gene automatic classification benchmark of the possible combinations between *Sc* and *Sc_B* with the PPI networks. Highlighted with * are the methods that can be applied to directed networks.

Algorithm	Sc	Sc+PPI_S	Sc+PPI_L	Sc_B	Sc_B + PPI_S	Sc_B+PPI_L
Louvain	0.051	0.054	0.080	0.070	0.117	0.083
Leiden	0.050	0.057	0.063	0.112	0.109	0.131
Leiden*	0.061	0.061	0.028	0.084	0.125	0.125
Walktrap	0.001	0.001	0.001	0.100	0.123	0.233
Walktrap*	0.001	0.001	0.002	0.099	0.130	0.226
Infomap	0.001	0.001	0.005	0.001	0.064	0.194
Infomap*	0.316	0.219	0.397	0.243	0.250	0.403

expression evidence and binding evidence. The latter is composed by the direct and more reliable associations that were documented in these networks and, for the largest network *Sc*, represents about 23% of the total number of links. We decided to create an additional network, *Sc_B*, using only these types of connections from the original *Sc* network to check if the results would improve. Another possible cause for the observed low results was the lack of sufficient documentation to build our networks, and consequently having incomplete biological information, so we additionally tested adding associations from protein-protein interaction (PPI) networks that had common genetic information with our yeast networks. We used two available PPI networks: the first from the CCSB Yeast Interactome Database[1] with 2 018 nodes and 2 930 interactions, which we will from now on call *PPI_S*, and the second a much larger network than the previous, *PPI_L*, with around 6 394 nodes and 994 296 connections, available in the STRING Database[2]. We added interactions that connected at least one node in common with the original *Sc* network. In Table 2, we can compare the results obtained previously for the original *Sc* network with the new *Sc_B* and with four larger networks that resulted from merging these two with the PPI's. From the new results we confirm our theories by observing that, generally speaking, the NMI metric scores improved when only considering the binding evidence for the creation of the *Sc_B* network and by adding the PPI's additional genetic information.

4 Multilayer Approach

A complementary approach to study the topology of the yeast networks is to apply a multilayer community detection algorithm, where each layer of the multilayer network is an individual yeast species and nodes from different layers can be connected. This approach allows us to cluster nodes from different species, which can be interesting since these species are biologically closely-related, so we might find groups of nodes with similar biological functionalities that belong to

1 http://interactome.dfci.harvard.edu/S_cerevisiae/.
2 https://string-db.org/.

different species and that end up being grouped up together due to these similarities. The algorithm that we will use in this section is Infomap [5], since it can be extended for multilayer networks and has already shown to be able to produce good partitions of our yeast networks in previous tests. We decided to use the *"two-level"* parameter/tag, meaning the core of the algorithm follows the Louvain method which can result in a fairly good clustering of the networks in a very short amount of time. In addition, there is already some work in the scientific community concerning random walks on multilayer biological networks [24].

Infomap allows the explicit addition of intra-layer (between nodes of the same layer) and inter-layer (between nodes of different layers) links. This is useful because we want to use mappings of nodes between each possible pair of yeast networks, available in YEASTRACT+, to create our multilayer networks. The approach we decided to follow was to create several multilayer networks with two layers each, where these two layers were assigned to each possible combination of yeast network pairs, and used Infomap to get the resultant communities of each of these multilayer networks. First, we wanted to check if the nodes that are connected through the inter-layer mappings are being well-clustered by the algorithm, which would otherwise show that these interactions between different species were not making a significant difference in the results, thus somewhat invalidating this alternative multilayer approach to our problem. In order to assess the legitimacy of this new approach, we created a metric that verified the percentage of nodes connected in the inter-layer interactions that were partitioned into the same community by Infomap. The results showed that almost all the mapped nodes were clustered together, with the large majority of the scores ranging between 90% and 100%. We also verified that Infomap was creating a very large number of communities for each multilayer, most partitions having more than 1 000 communities, with the exception of the multilayers that contained less documented networks.

Now that we know that the nodes linked between different species are being well-clustered, we can use the genetic classifications of the nodes to see if the multilayer communities are able to successfully group genes with similar functionalities, even if they belong to different species. A similar strategy to the one used when we individually evaluated the networks from a biological standpoint was used. We manually created communities using the nodes from both networks that compose the multilayer, where nodes with the same genetic classification or without a known one were added to the same community. Using these two community sets, created by Infomap and manually created by us, we used the NMI score function to compare the partitions and tell us if Infomap was able to successfully create communities where we could find groups of nodes with similar biological functionalities. In Fig. 1, we show these NMI scores for each pair of yeast networks represented through a heatmap so it is easier to read and analyze. We can see that the majority of the scores range between 20% and 40%, with the exceptions of the multilayers composed by smaller networks, such as *Zb* and *Km*, that are severely under-documented and consequently have less significant biological information contained in their structures.

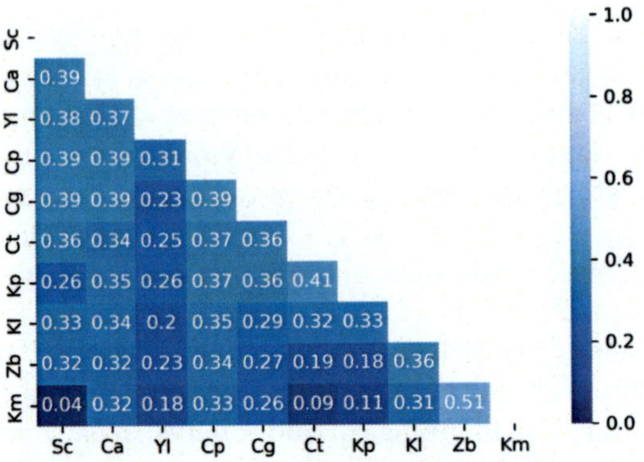

Fig. 1. Multilayer genetic evaluation heatmap with NMI scores for each pair of yeast networks, ordered by size.

The partitions retrieved contained a very large number of communities for each multilayer, but only a few with a significant amount of nodes. After a more detailed analysis of the *Sc* × *Ca* multilayer network, which contains the two more documented networks, we observed that only 6 communities from a total of 3 667 contained 10 or more nodes. In Fig. 2, we show the biological functionalities that were found in those 6 communities, as well as the percentage of nodes with the same function in relation to the total amount of nodes in each community. *C1* and *C5* are, by far, the communities with the largest representation regarding their size, both having more than 1 000 nodes, and regarding the variety of genetic functionalities of its genes, with 16 and 17 different genetic classifications, respectively. The first community is mostly made up of nodes from the *Sc* network, having only 2% of *Ca* nodes of a total of 1 436 nodes in this cluster, while *C5* nodes mainly belong to *Ca*. The remaining communities are much smaller and represented by one main biological functionality. If we look into the *C2* partition, we can see that it is composed by 13 genes, all with the same function: amino acid metabolism. The most interesting point here is that 7 of those nodes belong to the *Sc* network and the other 6 to *Ca*, which means that Infomap was able to group genes from different species with a similar biological functionality in their respective organisms. The same thing happens in *C3* where 21% of the nodes with the carbohydrate and energy metabolism function belong to *Ca* and, in the last community, 43% of the genes responsible for metal ion homeostasis belong to our largest network. We also analyzed the *Cp* × *Cg* multilayer network, and the first thing we observed was that although Infomap retrieved less communities than with the previous multilayer, there is a larger amount of communities with a significant amount of nodes in them. We saw a larger variety of genetic functionalities represented in its 7 largest communities, most of them present across every community, such as carbohydrate and energy metabolism which has the overall highest representation in these partitions. However, not all classifications are found across all partitions.

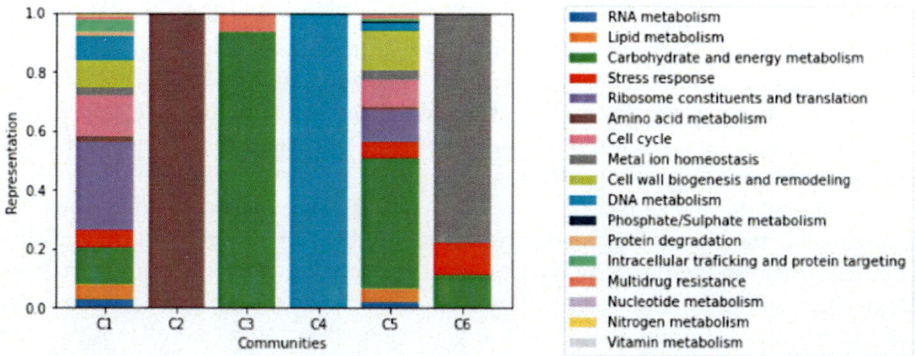

Fig. 2. Communities and respective genetic classifications for the *Sc* × *Ca* multilayer network. The bar of each term in a column represents the percentage of nodes with that classification in relation to the total number of nodes of the community. Only the 6 largest communities are shown.

For instance, the amino acid metabolism, phosphate/Sulphate metabolism and multidrug resistance functions are not found in a few communities. When comparing these partitions with the ones from the *Sc* × *Ca* multilayer, although we now have more communities with more genetic information, in each community we have less nodes with the same classification that belong to different species. Some communities are exclusively composed by nodes of one of the individual yeast networks. We can also see that the *Sc* × *Ca* multilayer contains genes with the vitamin metabolism function, which is not found in this multilayer.

5 Conclusion

Our approach in this paper allowed us to evaluate and compare a large group of community detection algorithms proposed by the scientific community and, consequently, to use the best of them to find significant communities in the yeast networks. We characterized and evaluated regulatory networks from these species regarding their biological functionalities, both from an individual analysis of each network as well as by using a multilayer approach. We accomplished this by applying the Infomap algorithm to multilayer networks composed by the links of two yeast networks and the mappings of associations between them. After analyzing some of the multilayer networks, we were able to identify functional structures conserved across species, since some of the communities contained nodes from different yeast networks with similar genetic classifications. All of this work contributed to the important study of transcriptional regulatory networks, thus increasing the existing knowledge in this scientific field. We conclude that studying the topological structure of these biological networks might not be enough to understand the underlying complex biological functionalities of these organisms, suggesting that more complementary information is needed to improve these results since, although transcriptional regulatory networks have been the subject of many studies in recent years, there is still a lot of unknown information about them that even biologists do not know.

As for the limitations, the first is that the choice of algorithms to partition the yeast networks was limited by their efficiency in running in a reasonable time and the available computational power to run these algorithms. Another possible limitation was that, although using the performance of the algorithms on benchmark networks is a better way to select them instead of going for popular choices, the tested benchmark networks might have different characteristics than the yeast networks, which can mislead us in the choice of the algorithms. Regarding the functional characterization chapter, the main limitation was that we did not have ground truth for the yeast networks, so as an alternative, we manually created this ground truth with some available knowledge of genetic classifications of some of the genes that compose these networks. In addition, labeling each node of the yeast networks with only one genetic classification can be reductive, since one gene can have several genetic functionalities in a species.

Finally, we leave here some suggestions for future work to be explored. To solve some of the limitations previously pointed out, labeling nodes with two or more classifications would be a good option, which could also facilitate the extension of the manually built partitions to allow overlapping. Furthermore, the inference of genetic information on the less documented networks using the available associations from the larger ones would also be interesting to study, thus also allowing the prediction of links in the smaller networks. Another complementary approach to the traditional detection of communities would be the discovery of motifs with significant representation in the identified modules, which could be particularly useful if applied alongside the Infomap partitions because this algorithm was able to find many structures with a small number of nodes that can be represented by these motifs.

Acknowledgements. This work was supported by national funds through Fundação para a Ciência e a Tecnologia (FCT) with references DSAIPA/AI/0033/2019 (Project LAIfeBlood), UIDB/50021/2020, UIDB/00408/2020 and UIDP/00408/2020 (INESC-ID and LASIGE multi-annual funding, respectively).

References

1. Babu, M.M., Luscombe, N.M., Aravind, L., Gerstein, M., Teichmann, S.A.: Structure and evolution of transcriptional regulatory networks. Current Opinion Struct. Biol. **14**(3), 283–291 (2004)
2. Barabási, A.-L.: Network science. Philosop. Trans. Royal Soc. A: Math. Phys. Eng. Sci. **371**(1987), 20120375 (2013)
3. Blondel, V.D., Guillaume, J.-L., Lambiotte, R., Lefebvre, E.: Fast unfolding of communities in large networks. J. Stat. Mech: Theory Exp. **2008**(10), P10008 (2008)
4. Danon, L., Diaz-Guilera, A., Duch, J., Arenas, A.: Comparing community structure identification. J. Stat. Mech: Theory Exp. **2005**(09), P09008 (2005)

5. De Domenico, M., Lancichinetti, A., Arenas, A., Rosvall, M.: Identifying modular flows on multilayer networks reveals highly overlapping organization in interconnected systems. Phys. Rev. X **5**(1), 011027 (2015)
6. Girvan, M., Newman, M.E.: Community structure in social and biological networks. Proc. Natl. Acad. Sci. **99**(12), 7821–7826 (2002)
7. Hagberg, A., Swart, P., Chult, D.D.: Exploring network structure, dynamics, and function using networkx. Technical report, Los Alamos National Lab.(LANL), Los Alamos, NM (United States) (2008)
8. Hubert, L., Arabie, P.: Comparing partitions. J. Classif. **2**(1), 193–218 (1985)
9. Lancichinetti, A., Fortunato, S.: Consensus clustering in complex networks. Sci. Rep. **2**(1), 1–7 (2012)
10. Lancichinetti, A., Fortunato, S., Radicchi, F.: Benchmark graphs for testing community detection algorithms. Phys. Rev. E **78**(4), 046110 (2008)
11. Lancichinetti, A., Radicchi, F., Ramasco, J.J.: Statistical significance of communities in networks. Phys. Rev. E **81**(4), 046110 (2010)
12. Latchman, D.S.: Transcription factors: an overview. Int. J. Biochem. Cell Biol. **29**(12), 1305–1312 (1997)
13. Lee, T.I., et al.: Transcriptional regulatory networks in saccharomyces cerevisiae. Science **298**(5594), 799–804 (2002)
14. Li, P.-Z., Huang, L., Wang, C.-D., Lai, J.-H.: EdMot: an edge enhancement approach for motif-aware community detection. In: Proceedings of the 25th ACM SIGKDD International Conference on Knowledge Discovery & Data Mining, pp. 479–487 (2019)
15. Monteiro, P.T., et al.: YEASTRACT+: a portal for cross-species comparative genomics of transcription regulation in yeasts. Nucleic Acids Res. **48**(D1), D642–D649 (2020)
16. Monteiro, P.T., Pedreira, T., Galocha, M., Teixeira, M.C., Chaouiya, C.: Assessing regulatory features of the current transcriptional network of saccharomyces cerevisiae. Sci. Rep. **10**(1), 1–11 (2020)
17. Newman, M.E., Girvan, M.: Finding and evaluating community structure in networks. Phys. Rev. E **69**(2), 026113 (2004)
18. Peixoto, T.P.: Descriptive vs. inferential community detection: pitfalls, myths and half-truths. arXiv preprint arXiv:2112.00183 (2021)
19. Raghavan, U.N., Albert, R., Kumara, S.: Near linear time algorithm to detect community structures in large-scale networks. Phys. Rev. E **76**(3), 036106 (2007)
20. Rossetti, G., Milli, L., Cazabet, R.: CDLIB: a python library to extract, compare and evaluate communities from complex networks. Appl. Netw. Sci. **4**(1), 1–26 (2019)
21. Rossetti, G., Pappalardo, L., Rinzivillo, S.: A novel approach to evaluate community detection algorithms on ground truth. In: Cherifi, H., Gonçalves, B., Menezes, R., Sinatra, R. (eds.) Complex Networks VII. SCI, vol. 644, pp. 133–144. Springer, Cham (2016). https://doi.org/10.1007/978-3-319-30569-1_10
22. Rosvall, M., Bergstrom, C.T.: Maps of random walks on complex networks reveal community structure. Proc. Natl. Acad. Sci. **105**(4), 1118–1123 (2008)
23. Traag, V.A., Waltman, L., Van Eck, N.J.: From Louvain to Leiden: guaranteeing well-connected communities. Sci. Rep. **9**(1), 1–12 (2019)
24. Valdeolivas, A., et al.: Random walk with restart on multiplex and heterogeneous biological networks. Bioinformatics **35**(3), 497–505 (2019)

Learned Monkeys: Emergent Properties of Deep Reinforcement Learning Generated Networks

Shosei Anegawa, Iris Ho, Khoa Ly, James Rounthwaite, and Theresa Migler(✉)

California Polytechnic State University, San Luis Obispo, CA 93405, USA
tmigler@calpoly.edu

Abstract. Graph generation methods used in the study of animal social networks suffer from the inability to fully explain individual agent behaviors and values. However, recent advancements in complex networks and machine learning offer a novel way of artificially simulating network formations. We use deep reinforcement learning (DRL) to model the proximity network of white-faced capuchin monkeys. This process of constructing a graph provides insight into the unique, individual social strategies of a network's agents depending on the initial DRL parameters. We generated a network to closely match the characteristics of the proximity network constructed from an observational dataset. For example, our model-generated graph and the observed graph consistently showed a few members with significantly higher betweenness centrality than all other members despite each agent starting with the same parameters.

Keywords: Animal social networks · Ecological network · Generative network models · White-faced capuchin monkeys · Deep reinforcement learning

1 Introduction

Our understanding of animal behaviors and animal socialization is underpinned in large part by our knowledge and analyses of their social networks. The study of animal social networks has largely relied on graphs, represented as complex networks of nodes and edges, constructed from data collected by field researchers or experimental observations. Analyses of simulations leveraging those constructed model networks can often provide great insight into the governing rules of social networks, but can struggle to illuminate individual behavior and decision-making processes.

Recent novel work in the field of complex networks has sought to apply deep reinforcement learning (DRL) - a paradigm of machine learning algorithms interested in simulated individuals, called agents, learning and accruing experience through interactions with their environment - to the construction of networks. The application of DRL in this context, referred to as building a network based on reinforcement learning, has produced networks with promising similarity to real network distributions [8].

A. S. Teixeira et al. (Eds.): CompleNet 2023, SPCOM, pp. 50–61, 2023.
https://doi.org/10.1007/978-3-031-28276-8_5

The contribution of this paper is an analysis of a dataset of Panamanian large brained white-faced capuchin monkeys from a graph theoretical standpoint, a presentation of the application of the metrics on the network as constructed from observational field data, and the application of a similar process of constructing a network based on reinforcement learning [6]. With continuous experimentation of our reinforcement learning model parameters, we matched the characteristics of our model-generated graph to that of the graph constructed from the observational data.

This paper is organized as follows: first, we discuss relevant work in order to highlight several recent applications of agent based modeling for studying animal social networks. Next, we offer a brief, non-exhaustive overview of reinforcement learning to provide adequate background for understanding our application of reinforcement learning in constructing the network. Subsequently, we present the graph metrics of our white-faced capuchin monkey dataset and our methods for building our network. Finally, we summarize our comparisons and findings between our generated network and the observed network.

2 Related Works

White-faced capuchins are not new to the concept of being analyzed via social networks. Crofoot et al. built a multitude of social networks from extensive observed data to learn more about white-face capuchins from a general social network theory perspective. This analysis has noted patterns of both aggressive as well as cooperative behaviors [1]. Perry et al. has also done extensive work with white-faced capuchins using social networks as a lens to learn more about their unique personalities and group structure [7], culminated as the Lomas Barbudal capuchin Monkey database and project [5]. More recently Perry and Smolla analyzed why these monkeys sometimes perform a complex and awkward series of social interactions coined as 'rituals' [6]. We pull proximity data from this paper as a foundation for our research. Lastly, the only major paper to combine agent-based modeling (ABM) with capuchin monkeys is Kajokaite's publication set to model male monkey dispersal movement. This ABM focuses on predicting specific actions of capuchin monkeys, whereas our model aims to model a proximity network [4].

There have been many studies using reinforcement learning to generate networks, although most have been used to characterize human social networks. Song et al. used a network generation method using reinforcement learning to represent a human social network. Their generated network was scale-free and small-world, characteristics seen in observed human networks. [8]. Also, work done which represents individual animals as their own agents to model group animal behavior has been found to be successful in studying grizzly bears by Hoegh et al. [2]. While these works used their representative models to predict group movement, the same underlying concept of representing each animal as its own autonomous intelligent individual has also been used to create a complete animal network [1]. Using models in this manner to generate networks is utilized in numerous other fields as well. Notably Jacobs et al. has presented a literature review of the field. Jacobs' work presents the numerous methods of generative networks and has provided sound techniques and optimizations to apply generative models from several perspectives [3]. Our paper uses both Song's and Jacob's work as a launch point for our model development.

The contributions of our work apply reinforcement learning as an agent-based network generation to analyze the unique and sociable white-faced capuchin monkeys.

3 Background

Reinforcement Learning. In this paper, we utilize reinforcement learning to model the behavior of capuchin monkeys. Reinforcement learning aims to have an agent maximize reward in the given environment. We define agents as simulated individuals that interact with and learn from the environment and other agents through actions and reward. A reward is a function that maps states of the environment to a positive or negative value. An agent takes actions moving them from different states seeking maximal value. The value is then defined as a long-term reward, or the sum of all the rewards from the initial state to the terminating state. Given a state, an agent will determine the optimal action based on the policy π, which is a function that maps states to actions based on the reward. The goal is to then have an optimal policy, whereby given any state, the agent provides the action that maximizes reward. For the most part, the policy heavily relies on the value function, which is what the agents need to learn. The value function represents the long-term reward given the current policy, and thus desirability, of any given state.

We employ an algorithm known as *Q-learning* for our agents. Q-learning aims to produce a Q-Function which is an extended definition of a value function. Unlike the traditional value function that takes in a state and returns a value given the current policy, the Q-function takes in a state-action pair. $Q_\pi(S, A)$ is defined as the value of taking action A from state S, and then thereafter following the current policy π. In addition to this, Q-learning is known as an online reinforcement learning algorithm, meaning the Q-function is updated during the sampling process, resulting in a dynamic policy that improves throughout an episode. A single update step is defined as follows [9]:

$$Q(S_t, A_t) \leftarrow Q(S_t, A_t) + \alpha[R_t + \gamma \max_a Q(S_{t+1}, a) - Q(S_t, A_t)] \tag{1}$$

where α is the learning rate, A represents the set of all actions, S is the set of all states, and R represents the immediate reward assigned. γ represents the discount factor, which decreases the value of a reward the farther it is in the future. A bigger α value means that the Q-function is updated in larger steps, which can help the value function converge more quickly, but can also harm convergence if set too high. Our model uses a neural network to approximate this function, with its loss defined as the mean squared error of [9]:

$$(R_t + \gamma\, max_a\, Q(S_{t+1}, a)) - Q(S_t, A_t)$$

In this case, we aim to get the current prediction of $Q(S_t, A_t)$ closer to $R + \gamma * max_a(Q(S_{t+1}, a))$, which is at least as accurate as $Q(S_t, A_t)$, and serves to bring the current approximate Q-function closer to the correct Q-function.

This then leads us to the policy, which as mentioned previously is determined by the Q-function. In our case, we employ an ϵ-greedy policy. This policy normally chooses actions greedily by using the max_a over actions of $Q(S, A)$, but with a probability of ϵ takes a random action. This exists so that the agent can balance exploring new options and exploiting existing options. Tuning ϵ is an extremely important step in any Q-learning model.

Slowly the agent learns more accurate values in the Q-function and is able to take actions that yield higher rewards. However, the introduction of an ϵ-greedy policy forces the agent to take possibly worse actions, but occasionally this exploration results in new states with greater rewards. This tradeoff of exploring new paths and staying with what is known to work propels Q-learning forward.

4 Observed Dataset

Dataset and Proximity Network Model. The data relevant to this paper comes from Perry et al. and the Lomas Barbudal Capuchin Monkey database [5]. More specifically, we will be concerned with the data detailing the number of interactions per dyad of monkeys from the 'Flakes' group observations. The data includes 47 individuals, 761 dyad pairs, and 26797 proximity observations over a multi-year period between 1 February 2004 and 11 October 2018. In this network vertices represent white-faced capuchin monkeys and two monkeys are connected if they were observed within 40 cm of each other. The edges are weighted with an SRI metric, a measure of total proximity observations between dyad pairs normalized by the total scans per individual monkey in the dyad. The resulting network had degrees ranging from 5 to 44 with the mean degree being 26.9. The network had an average path length of 3.615 and an average clustering coefficient of 0.831 (Figs. 1, 2, 3 and 5).

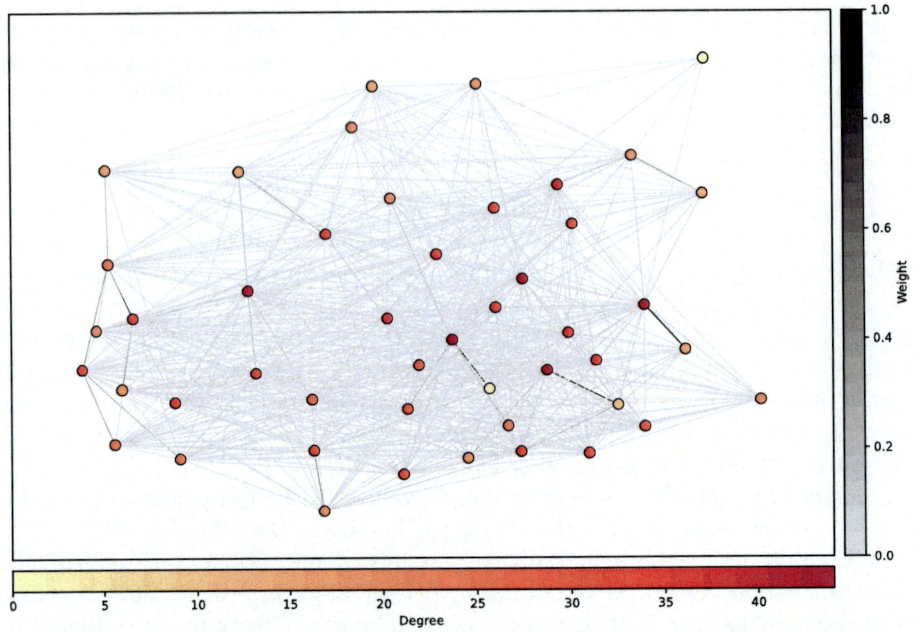

Fig. 1. The proximity graph (47 nodes) of the observed white-faced capuchin dataset with weight defined as SRI

5 Reinforcement Learning Model

Following Song et al.ś recent success, we use their general architecture as the foundation for our model. [8]. We define a space ranging from 7×7 to 10×10 states on a board, and place 47 agents in it in order to model our data. Each agent has their own independent Q-function approximator, which are each 3-layer neural networks with 128, 128, and 64 units respectively, eventually returning a number representing the value. During graph formation, we randomly initialize the weights to the Q-function and the positions of each individual on the board, and build an initial graph based on this. At each timestep, an individual can take one of five actions: move up, down, left, right or stay. If two or more individuals end up in the same state, the weights on the graph are incremented by a value ω. If two individuals have a weight greater than zero and are not in the same state in a given timestep, their weights are decremented by a value τ. If at this point, the weights for an edge reach zero, the edge is removed. After the weights are updated at the end of a timestep, rewards are distributed to agents equal to the total of their current weights. This pushes them to both maintain and form new social connections throughout an episode.

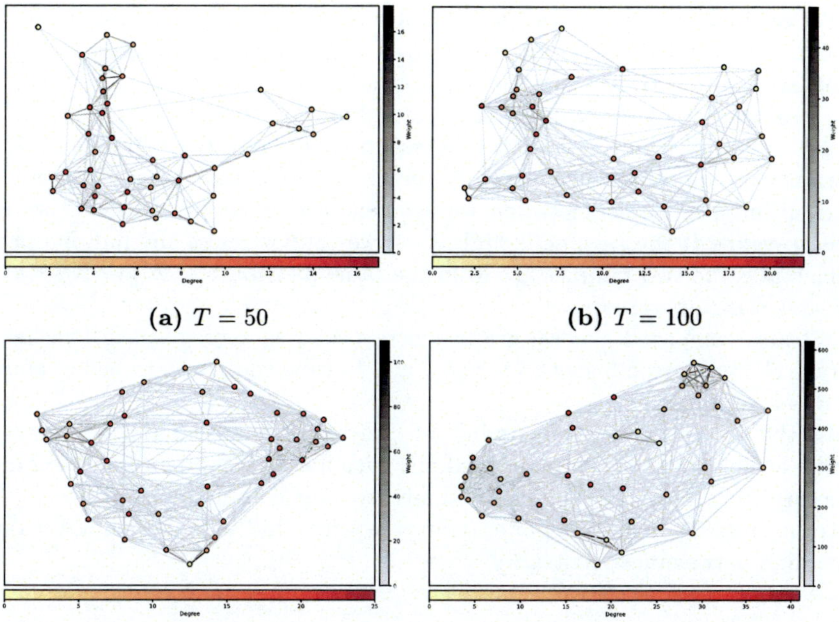

(a) $T = 50$ (b) $T = 100$

(c) $T = 250$ (d) $T = 1500$

Fig. 2. As time progresses the generated network more closely resembles the observed graph

6 Methods

Model Tuning—Episodes and Iterations. The first thing we tuned for our model was the decay of weights on the graphs over time, τ. We found that this had the greatest impact on the KS-tests, clustering coefficient, and average path length. We settled on a decay $\tau = -0.005$ per timestep, relative to $+1$ reward for forming or maintaining connections. We hypothesize this microscopic decay performed the best due to the observed graph being a proximity graph, where connections cannot be removed once formed. In addition, we tune ϵ which represents the inclination of a monkey to explore new connections or to focus on retaining current connections.

Finally, we tune γ, the discount factor, which represents to what extent monkeys take into account future reward, or how far they look ahead into the future to make decisions.

Generative Network Comparison. To test how well our generated graphs matched characteristics of the original graph, we created a baseline model and compared iterations of the RL model and the baseline model to the observed graph. The baseline model has agents taking random actions at every timestep.

We utilized various metrics to show the validity of our generated network. Focusing on just the degree distribution of the generated and observed graph, we utilize a couple of methods to compare equality.

The first is the two-sample Kolmogorov-Smirnov (KS) test, which produces the KS statistic. The KS statistic represents how likely the two sets of degree distributions could have been drawn from the same unknown degree population distribution. The implementation we use generates a KS statistic and an associated p-value. If the p-value is high, then we cannot reject the null hypothesis which states the two samples are from the same distribution. A low KS statistic indicates a high p-value.

We also compared general graph metrics such as average path lengths and average clustering coefficients. Our goal was to tune the graph to better emulate the observed graph.

Lastly, we compare the number of "keystone" individuals in all graphs. Krause et al. defines keystone individuals with a strong influence on the graph's structure and group. Keystone individuals are most commonly identified through centrality metrics [10]. This paper defines keystone individuals as nodes exhibiting a high betweenness centrality.

7 Results

Model Comparison. For most of this section, we fixed ϵ at 0.3, the discount factor γ at 0.9, and the board size at 10×10. The graphs are flattened over 50 iterations of 1500 timesteps episodes. As a baseline for showing model improvement we compared the graphs produced by our agents to a second model where agents took random actions at every timestep. We found that most of the metrics that we used tended to converge after the model had ran for 1500 timesteps, so all graphs shown represent the progress of the model leading up to 1500 timesteps. In the following section we will present each metric used to compare our generated graphs to the real-world proximity graphs of the capuchin monkeys. We also experimented with board sizes smaller and larger than 10×10, but over 1500 timesteps, this made little difference.

Degree Distribution Comparison. Degree distribution is an important metric for a social network as it gives deep insight into the structure of a network. In this section, by comparing the degree distributions of our generated graphs to that of the observed graphs, we show that our model generates graphs representative of the real world. With the above parameters, the random acting agent performed significantly worse in the KS test. The random acting agent model reached its best metric for the two-sample KS test at around 200 iterations, with a KS-test statistic of around 0.38 and a p-value of 0.36. On the other hand, the RL model reaches a KS-test statistic of 0.17 and a P-value of 0.77. It should also be noted that at low timesteps such as 200 iterations, the graphs tend to be more sparse which makes these tests less reliable. This shows that the degree distribution of our model was significantly more closely aligned to the proximity graph's degree distribution. With a significance level of 0.77, the RL generated a degree distribution similar enough to the observed degree distribution that the two could have come from the same population.

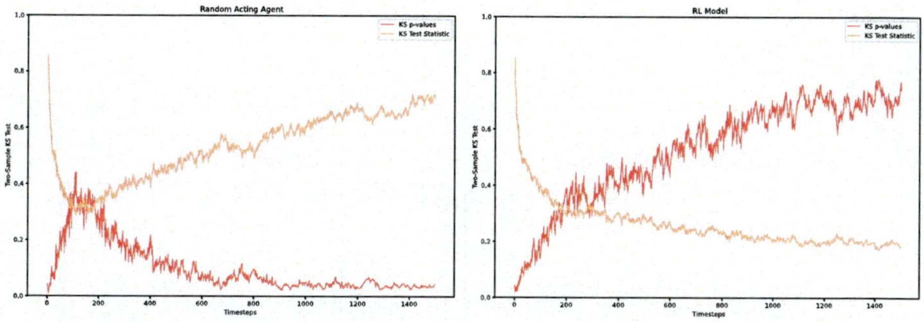

(a) Random Acting Agent **(b)** RL Model

Fig. 3. Throughout an episode, Fig. 3b has consistently decreasing KS test statistics and increasing p-values, highlighting similarity between our RL model and the observed proximity network.

It is also interesting to note that both the RL model and random acting agent have similar results up to around 200 timesteps. This makes sense, considering that our Q-function approximator's learning rate parameter is set to a fixed 0.01, so meaningful learning should begin being reflected around after 100-200 updates.

Keystone Individuals/Centrality Comparison. The second metric we used was the betweenness centrality of each individual member of the graph. In the observed proximity graph, there are two "keystone" individuals, who had significantly higher betweenness centrality than all other members in the graph and were hubs for social interaction. Interestingly, the graph generated by the RL model often had 2-3 keystone individuals with significantly higher betweenness centrality over others. This means at each iteration some agent-policy optimizations always led towards having 1 or 2 keystone individuals. Shown in Fig. 4 centrality over each individual agent/monkey. The similarities in the number of keystone individuals is striking, because it shows how agents learn to build connections in a similar fashion as monkeys in the observed data. Despite all agents being initialized with the same Q-function and parameters, we still observed consistent differences between individual agents in a similar manner to our observed proximity graphs. It was notable to see that even over 50 distinct episodes of 1500 timesteps, a few agents consistently became important within the group of monkeys.

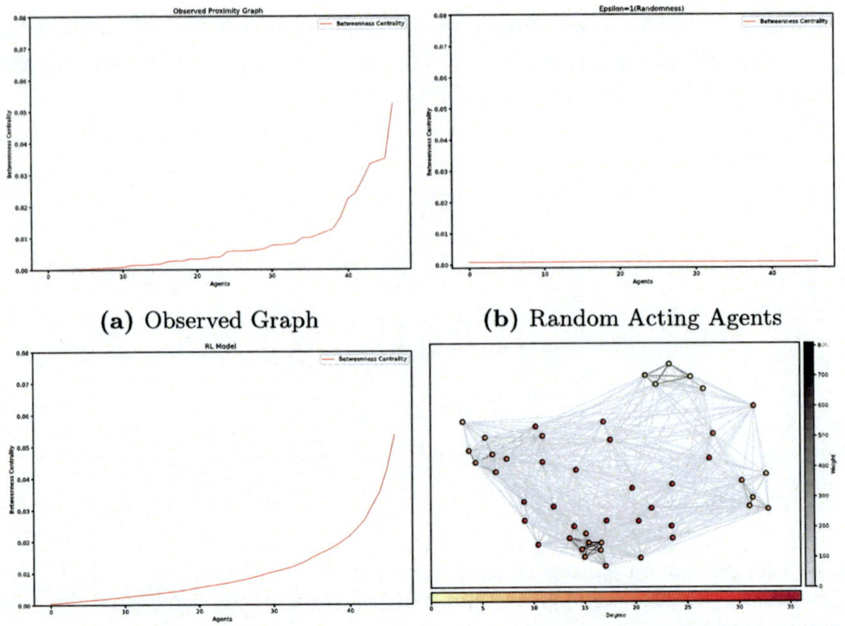

(a) Observed Graph **(b)** Random Acting Agents

(c) RL Model **(d)** Sample Graph

Fig. 4. Figure 4b had most nodes with extremely low betweenness centrality, with each node having the same connections around 1e-5, due to the graph having minimal connections between clusters. The smoothness of the Fig. 4c graph results from it being the average over 50 episodes. In addition to this, Fig. 4d clearly shows the individuals that are keystone, with very central positions on the graph. In this case, Agents 1 and 11 have the highest centralities

Clustering Coefficient and Average Path Length. In general, the random agent graphs generated had a strong clustering coefficient, almost reaching 1.0 over 1500 timesteps, and a lower average path length, at around 3.8. This number might be disproportionately low because given that the graph is relatively sparse, nonexistent paths are not factored in the average path length metric. The RL generated network ended with a clustering coefficient of 0.73 and an average path length of 4.5. Recall that the observed proximity graph had an average clustering coefficient of 0.81, and an average path length of 3.7. From these we see that the RL graph better modeled the observed data on clustering coefficient, but not in average path length.

The RL graph generation method outperforms the random acting agent model at every metric aside from average path length when attempting to represent how monkeys act in nature.

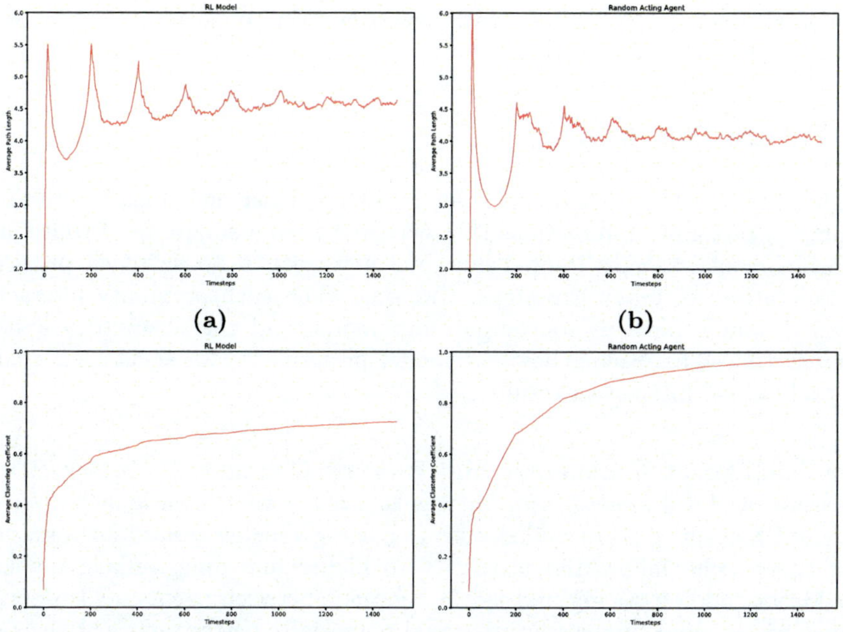

(a) (b) (c) (d)

Fig. 5. The random acting agent has a more similar average path length to the observed dataset, while the RL model is more similar in clustering coefficient

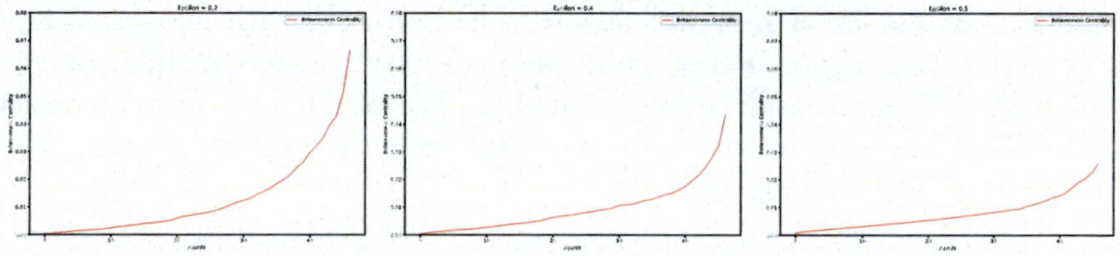

(a) $\epsilon = 0.2$ **(b)** $\epsilon = 0.4$ **(c)** $\epsilon = 0.5$

Fig. 6. ϵ has a high impact on the existence of clear keystone individuals in the generated graphs. The relative roughness of these graphs is due to them being averaged over a smaller number of episodes

Epsilon. Next we varied ϵ, the coefficient of exploration. The metric most correlated with changes in ϵ was the distribution of betweenness centrality, and thus keystone individuals throughout the network. Some significant increments are shown below in Fig. 6. In general, as ϵ grows larger, the more prominent keystone individuals become. As it approaches 1, individuals begin to have very evenly distributed connections, and thus behave like a single large cluster rather than a collection of smaller cliques. ϵ is a metric of how exploratory our agents are. Due to the extreme similarity of the centrality of our graph generated with $\epsilon = 0.3$, we can hypothesize that out of every 10 interactions that these monkeys have on average, 3 will likely be with new or weaker connected monkeys, while 7 will be with closer connections.

8 Future Directions

Limitations of the Research. Our current dataset only includes the total number of proximity interactions that occurred from a subgroup of white-faced capuchin monkeys. This limited our final networks to be flattened proximity representations of many timesteps. Knowing when each proximity interaction occurred would allow us to compare our model and the observed graphs at different timesteps. Seeing how the model performs as its agents learn could yield additional interesting results.

Increased Agent Uniqueness and Reward Changes. To better simulate the behavior of white-faced capuchin monkeys in the wild, additional parameters for each RL agent could be added, such as placing a higher reward for connecting with agents who fall within proximity to higher matriline ranking monkeys. In addition, adding sibling and family relations between agents, or sex-specific characteristics, and tuning rewards could potentially better represent the social structure of the capuchin monkeys.

Social Interactions and Network. Adding in social interactions opens a wide range of research opportunities to utilize game theoretics, treating social interactions as games of risk and reward. Furthermore, with enough complexity, we could even model Perry's and Smolla's observations of intricate ritual interactions and relationship quality index [6]. We could also apply our method of generating a network to other primates to see if the patterns persist beyond white faced capuchin monkeys.

9 Conclusion

We modeled white-faced capuchin monkeys as reinforcement learning agents using an ϵ-greedy policy to generate proximity networks. Running our model to generate flattened graphs over 50 iterations of 1500 timestep episodes, we found our graphs had many similar properties to a real-world proximity network. The most prominent result we observed was the consistent presence of a small number of keystone individuals in our generated graphs. We also observed how the parameter ϵ impacted the distribution of keystone individuals. Furthermore, in comparison to a simplistic graph generated by random acting agents, our network's degree distribution was more consistently and closely related to the observed graph.

Acknowledgements. We would like to express our deep gratitude to Dr. Susan Perry, for her enthusiasm in our work and dedication to studying white faced capuchin monkeys. We would also like to thank Colin Chun, Evan Diaz, Spencer Wong, and Allen Yu for their camaraderie during our initial exploration of animal networks. Lastly our greatest thanks for the support of Linda Anegawa and Robert Rounthwaite whose help propelled this paper to new heights.

References

1. Crofoot, M., Rubenstein, D., Maiya, A., Berger-Wolf, T.: Aggression, grooming and group-level cooperation in white-faced capuchins (cebus capucinus): Insights from social networks. Am. J. Primatol. **73**, 821–33 (2011). https://doi.org/10.1002/ajp.20959
2. Hoegh, A., van Manen, F.T., Haroldson, M.: Agent-based models for collective animal movement: proximity-induced state switching. J. Agric. Biol. Environ. Stat. **26**(4), 560–579 (2021). https://doi.org/10.1007/s13253-021-00456-0
3. Jacobs, A.Z., Clauset, A.: A unified view of generative models for networks: models, methods, opportunities, and challenges (2014). https://doi.org/10.48550/ARXIV.1411.4070. https://arxiv.org/abs/1411.4070
4. Kajokaite, K.: Male dispersal decisions: an agent-based general model and suggested refinements for white-faced capuchin monkeys (cebus capucinus) (2014)
5. Perry, S., Godoy, I., Lammers, W.: The lomas barbudal monkey project: two decades of research on cebus capucinus, pp. 141–163 (2012). https://doi.org/10.1007/978-3-642-22514-7_7
6. Perry, S., Smolla, M.: Capuchin monkey rituals: an interdisciplinary study of form and function (2020). https://doi.org/10.1101/2020.02.21.958223
7. Perry, S.E., et al.: Social conventions in wild white-faced capuchin monkeys. Curr. Anthropol. **44**, 241–268 (2003). https://doi.org/10.1086/345825
8. Song, W., Sheng, W., Li, D., Wu, C., Ma, J.: Modeling complex networks based on deep reinforcement learning. Front. Phys. **9**, 822581 (2022). https://doi.org/10.3389/fphy.2021.822581. https://www.frontiersin.org/articles/10.3389/fphy.2021.822581
9. Sutton, R.S., Barto, A.G.: Reinforcement learning: an introduction, pp. 135–181. The MIT Press, second edn. (2018)
10. Wilson, A.D.M., Krause, J.: Personality and social network analysis in animals. In: Animal Social Networks. Oxford University Press (2015). https://doi.org/10.1093/acprof:oso/9780199679041.003.0006

Targeted Attacks Based on Networks Component Structure

Issa Moussa Diop[1](✉), Chantal Cherifi[2], Cherif Diallo[1], and Hocine Cherifi[3]

[1] LACCA Lab, Gaston Berger University, Saint-Louis PB 234,, Senegal
{diop.issa-moussa,cherif.diallo}@ugb.edu.sn
[2] DISP Lab, University of Lyon 2, Lyon, France
chantal.bonnercherifi@univ-lyon2.fr
[3] ICB UMR 6303, CNRS, University Bourgogne Franche-Comté, Dijon, France
hocine.cherifi@u-bourgogne.fr

Abstract. Robustness analysis for targeted attacks is essential, especially for critical infrastructures. Typically, targeted attacks rank the nodes according to a centrality measure and remove top nodes according to a budget. The goal is to exploit the network features efficiently to dismantle them with a minimal budget. Few works are linked to the network mesoscopic properties in the literature, although it is well-admitted that communities or core-periphery are ubiquitous in real-world networks. We propose a network dismantling method based on a mesoscopic representation called the component structure. It performs classical centrality attacks on the network's global components rather than on the original network. Global components of a network are isolated networks formed by the interactions between its dense parts that one can extract from a community or multiple core-periphery structures. We investigate the proposed strategy using three real-world networks and popular centrality measures (Degree, Betweenness, and PageRank). Results show that the proposed approach is more effective for Degree and PageRank. In contrast, the Betweenness attack on the original network slightly outperforms the attack on the global components but at the price of higher complexity.

Keywords: Robustness · Component structure · Network resilience

1 Introduction

Robustness analysis is one of the research areas with significant interest in the network science literature. Many networks are susceptible to failure or attack, especially infrastructure networks, whose damage can affect society at different levels. Studying the robustness of a system consists in evaluating its vulnerability against failures or intentional attacks. Thus, many articles study or propose attack strategies, mainly based on centrality measures [1–4]. However, few consider the influence of the mesoscopic organization of the network on its robustness. We briefly discuss the main contributions in that direction considering the network community structure in the network dismantling process.

A. S. Teixeira et al. (Eds.): CompleNet 2023, SPCOM, pp. 62–73, 2023.
https://doi.org/10.1007/978-3-031-28276-8_6

In [5], the authors propose link and node-based frameworks exploiting the community structure to dismantle a network. First, one needs to uncover the network community structure. In their experimental study, the authors evaluate five community detection algorithms (Louvain, Girvan-Newman, Clauset-Newman, Label Propagation, and Fluid community). Then, they build the community network considering each community as a node and establishing a link if two communities share at least a connection. The dismantling strategy aims at disconnecting the communities. Therefore, they attack links in the condensed community network and nodes to dismantle the original network. Accordingly, they select critical links in the reduced community network and map them into nodes in the original network. According to the preceding step, a measure of importance to target a node or link is designed relying on five criteria. The last level is about translating the attacks on the condensed network to the initial network. They obtain up to 40 community-based network dismantling methods. They compare these methods to 7 classical network dismantled strategies. R [] is the evaluation criteria. The experiments include real-world and artificial networks. Result show Community based dismantling is sensitive to the community detection algorithm. In addition, community-based network dismantling is not efficient for model networks. In most real networks, the proposed methods are generally more effective than the alternatives and are also much more efficient.

In [6], the authors present a dismantling network framework based on community structure. Their method is an iterative procedure in which they remove the node with the highest inter-community links in the community with the largest size. In each iteration, the community detection Leiden algorithm is used. They investigate three real networks and a random network. The percolation threshold is computed to compare their method to the classical degree attack. They find that their strategy outperforms slightly on the random networks. For the real-world network, the community dismantling is more effective than the degree centrality dismantling strategy. Nevertheless, the efficiency of their method decreases when the community structure strength decrease. In addition, the complexity of the method is very high.

To summarize, these works demonstrate the advantages of exploiting the mesoscopic representations of real-world networks to design effective and efficient attack strategies. Our study is in this direction. Indeed, we propose and investigate a dismantling technique based on a new mesoscopic structure described in [7]. The component structure of a network splits networks into two types of components. The local components are the dense parts of the network. The global components contain the nodes and the links joining the local components in the original network. In previous work, we show how this new representation allows a better understanding of the local impact of various classical centrality-based attacks on network robustness [8,9]. Here we present a dismantling strategy based on a network's global components. With the global components, one can see, for example, that the inter-community links of a network can form several isolated networks. The purpose is to attack the global components using a given centrality-based strategy instead of the overall network and to compare with the corresponding attack on the original network. We investigate three real-world networks with different community structure strengths in the experimental evaluation. Results show that the proposed framework outperforms most classical attacks. It is also more efficient.

2 Background

2.1 Component Structure

The density of real-world networks is generally not uniform. One usually captures this phenomenon using two mesoscopic features: 1) the community structure and 2) the core-periphery structure. Although there is no consensus on a universal definition of these representations, they share that the network contains groups of nodes tightly connected, called cores or communities. They are supposed to be loosely related to other groups when considering the community structure. Peripheral nodes sharing few connections surround these core groups in the multi-core-periphery structure. The component structure builds in these representations. It splits the networks into dense groups and their interactions. One obtains the local components by isolating the dense parts of the networks. Links and nodes connecting the local components form the global components. To build the component structure, one proceeds as follows:

To build the component structure, one proceeds as follows:

1. Uncover the dense parts of the network.
2. Remove the links between the dense parts to extract the local components.
3. Remove the links within the dense parts and the subsequently isolated nodes to extract the global components.

Note that this representation is redundant. Indeed, a node can simultaneously belong to a local and a global component. One can use community detection or multi-core-periphery algorithms to extract the dense parts of the network. In this work, we consider an approach based on the community structure to uncover the component structure. Figure 1 A describes the extraction process of the component structure. In this example, one uses a non-overlapping community detection algorithm to extract the dense parts of the network. Then, we form the local components by removing the inter-community links. Removing the intra-community links and the isolated nodes extracts the global components.

2.2 Targeted Attack

Targeted attacks aim to remove the most vital nodes for network connectivity [10–12]. Centrality measures generally describe the importance of nodes [13,14]. In a strong attack strategy, one removes nodes in the network in descending order of magnitude of the chosen centrality. Classically, one can distinguish three types of centrality measures: Neighborhood-based, Path-based, and iterative refinement [15]. This work uses the most popular measures in each category: Degree, Betweenness, and PageRank.

Degree is a centrality based on the neighborhood [15]. In other words the node influence is computed based on its local neighbors. Indeed, given a graph $G(V, E)$, such as V is the set of nodes and E the set of links, the Degree k_i is the number of the direct neighbors of a node i. It has a local scope, and is defined as:

$$k_i = \sum_{j \in V, i \neq j} a_{ij}$$

a_{ij} is an element of the binary adjacency matrix of G such as $a_{ij} = 1$ if i and j are connected, otherwise, $a_{ij} = 0$.

Betweenness is a global centrality based on path [15]. The fraction of the shortest path passing through a node i is its Betweenness. When it is normalized, the Betweenness of the node i is defined as:

$$b(i) = \frac{2}{(n-1)(n-2)} \sum_{i \neq j} \frac{\sigma_{jk(i)}}{\sigma_{jk}}$$

σ_{jk} is the number of the shortest path between j and k. $\sigma_{jk}(i)$ is the number of the shortest path from j to k passing in i.

PageRank is a centrality based on iterative refinement [15]. Indeed, the influence of node depend on the influence of its neighbors which in turn also depends of their neighbors. Initially defined for directed networks, the PageRank in undirected networks consider two directions for a link. At t step, the PageRank of a node i is defined as follows:

$$pr_i(t) = \sum_{j=1}^{n} a_{ji} \frac{pr_j(t-1)}{k_j^{out}} \tag{1}$$

k_j^{out} is the out-degree of the node j.

2.3 Evaluation Measures

We use the Largest Connected Component(LCC) and the R-value to evaluate the network's resilience. During the process of removing nodes from a network, the largest interconnected set of nodes is called the Largest Connected Component. The larger the LCC, the less effective the attack on the network. The R-value refers to the size of the LCC during the process of removing nodes [16]. R is defined as follows:

$$R = \frac{1}{N} \sum_{N_r=1}^{N} s(N_r)$$

$s(N_r)$ is the size of the LCC after N_r nodes are removed. R represents the area under the curve that denotes the LCC progression when a node is removed. It ranges between $\frac{1}{N}$ and 0.5. The smaller the R and the number of nodes required to break up the network, the more effective the attack.

3 Data and Methods

3.1 Data

We use three real-world networks (infrastructure, social, and information networks) to conduct our experiment. The infrastructure network concerns the Brazil bus network [17]. A node represents a bus station in a municipality, and there is an edge between two nodes if the bus stations share at least one route. The nodes in the social network are Facebook's pages denoting Public Figures. A link is established when there is a mutual like between them [18]. The information network is a co-author network [19]. The nodes represent the author, and a link means two authors appear at least once in the same paper. These three networks are undirected and unweighted. Their basic topological properties are reported in Table 1.

Table 1. Basic topological properties of the networks under study. N is the network size. $|E|$ is the number of edges. d is the diameter. l is the average shortest path length. ν is the density. ζ is the transitivity also called global clustering coefficient. $k_{nn}(k)$ is the assortativity also called Degree correlation coefficient.

Networks	N	$\|E\|$	d	l	ν	ζ	$k_{nn}(k)$
Brazil bus	1786	19060	6	2,81	0,01	0,21	−0,01
Public figures	11565	67038	15	4.62	0.001	0.16	0.2
Co-author	13861	44619	18	6.27	0.0005	0.35	0.157

3.2 Methods

The proposed method relies on the component structure. First, one separates a network into local and global components. Suppose we refer to the well-known community structure used to uncover the dense parts of the original network. In that case, the local components contain the nodes in a community and their related intra-community links. Therefore, there are as many local components as communities. The global components include the nodes sharing links with the other communities and their inter-community links. As the components are isolated networks, one can perform a targeted attack on any of them rather than the original network. We propose dismantling the network by performing a targeted attack on its global components. Indeed, removing nodes in the global components allows for the isolation of the local components. Note that the proposed strategy is generic. Indeed, one can use any centrality measures or any method available to fragment a network as long as it targets the global components. After uncovering the component structure, the process proceeds as follows:

1. Rank the global components in descending order of size
2. From the largest to the smallest global component
3. **do**
4. Rank the nodes of the global component according to a centrality measure
5. Disconnect the top nodes of the global component.
6. Disconnect the same nodes in the original network.
7. Extract the LCC in the original network.
8. Extract the LCC of the global component.
9. **While there is a link in the LCC of the global component**

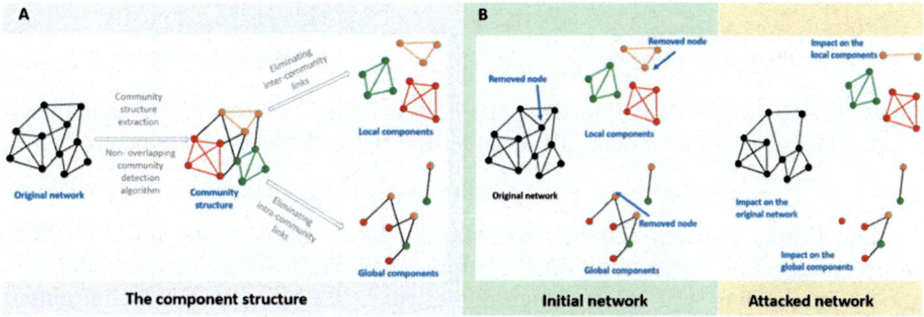

Fig. 1. (A) Process to uncover the component structure. We use a community detection algorithm to uncover the dense parts of the network in this example. Therefore the 3 communities are the 3 local components and there are 2 global components. (B) Attack on the largest global component and its impact on the original network.

4 Experimental Results

4.1 Component Structure

We use the Louvain community detection algorithm to uncover the community structure. Then, we extract the component structure of each network. In the following, we present the component structure of each network. The Brazil Bus network consists of 9 local components and three global components. The local components correspond to limited geographical areas. Note that these areas overlap. Among the local components, the smallest contains ten nodes (less than 1% of the overall network). The most significant local component includes 22.6% of the nodes of the network (404 nodes). Most of the diameters of the large local components are identical to those of the initial network. The smallest diameter is 4. The large local components tend to be more transitive than the Brazil Bus network, except the largest, which is the least transitive.

All the local components are more disassortative than the initial network. In the global components, the largest contains 53.24% of the entire network (951 nodes), while the two others have 3 and 2 nodes.

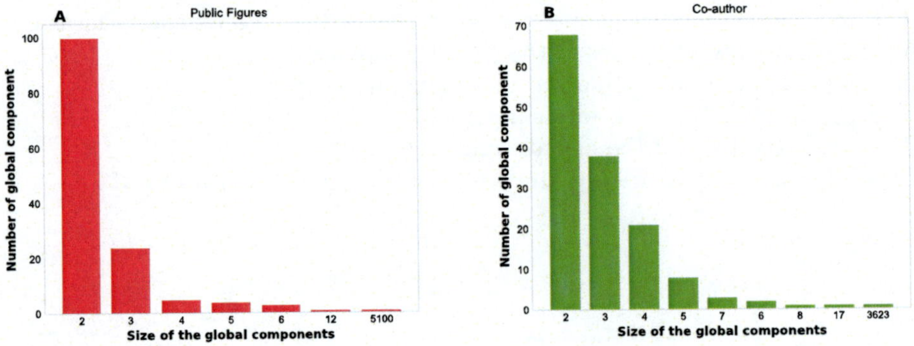

Fig. 2. (Left) Distribution of the size of the Public Figures network global components. (Right) Distribution of the size of the Co-author network global components.

The Public Figures network contains 34 local components and 138 global components. The local components include 19 large local components, with almost more than 1% of the original network. The largest local components size range between 1.21% and 16.8%. The diameter of the local components size range between 7 and 20. Only two large local components have a larger diameter than the initial network; most measure 11 or 12. Generally, the large local components are less transitive, except for a few. Their transitivity measure range between 0.07 and 0.51. In contrast to many social networks, 10 of these large local components of the Public Figures network are disassortative.

Figure 2A reports the distribution of the size of the global components of the Public Figures network. The largest global component includes 44% of the nodes. All the other global components are small. The largest global component requires 16 hops at maximum to join two nodes, one more than the original network. Moreover, it is less transitive and disassortative.

The Co-author network contains 62 local components and 143 global components. The largest local component has 4% (555 nodes) of the nodes of the overall network. The two small local components have less than 1% of the nodes. The smallest among the local components includes less than 0.64%. Only one local component has the same diameter as the original network. The smallest diameter among the other local component is 8. Except for four large local components, the other components are more transitive than the Co-author network. In contrast to the initial network, a third of the large local components are not assortative.

Figure 2B illustrates the distribution of the size of the global components of the Co-author network. Its diameter is one more hop greater than the one of the initial network. Nevertheless, it is by far less transitive and disassortative. The largest global component contains 26% (3623 nodes) of the nodes of the overall network. The smallest ones have two nodes.

To summarize, let us focus on the global components of each network. Indeed, we perform attacks on these components. One can see that each network has a large global component and several small global components.

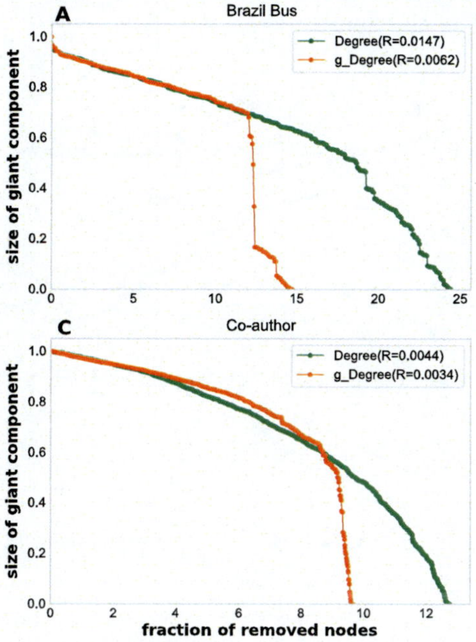

Fig. 3. Degree-based attacks: Evolution of the relative size of the LCC of the original network when the attack is performed on the original network (in green) and on its global components (in orange) A) Brazil Bus network B) Public Figures network C) Co-author network. The figures contain also the R values.

4.2 Attacks Evaluation

Degree Attacks. Figure 3 shows the evolution of the relative size of the LCC as a function of the proportion of top-degree nodes removed in the global components in the three networks under study. It also reports the same curves when performing the degree attack strategy in the original network for comparative purposes. One can see that the proposed attack on the global component is far more efficient than the classical attack on the overall network for all the networks. Indeed, attacking the global components of these networks requires fewer nodes to dismantle the three networks. In addition, the R values are smaller compared to the degree-based attack on the entire network.

Nevertheless, one can see in Fig. 3 that one needs to reach a certain proportion of removed nodes before observing high differences between the two strategies. Indeed, removing less than 12% of the nodes in the Bus Brazil network produces similar damage for both attacks. Beyond this value, the LCC decreases sharply, with the proposed attack strategy showing its superiority. The same observation holds in the two other networks. Indeed, for the Public Figures network, one needs to reach a proportion of 15% of removed nodes before observing substantial differences. The global component attack becomes more effective in the Co-author network when removing 4 to 6% of the nodes.

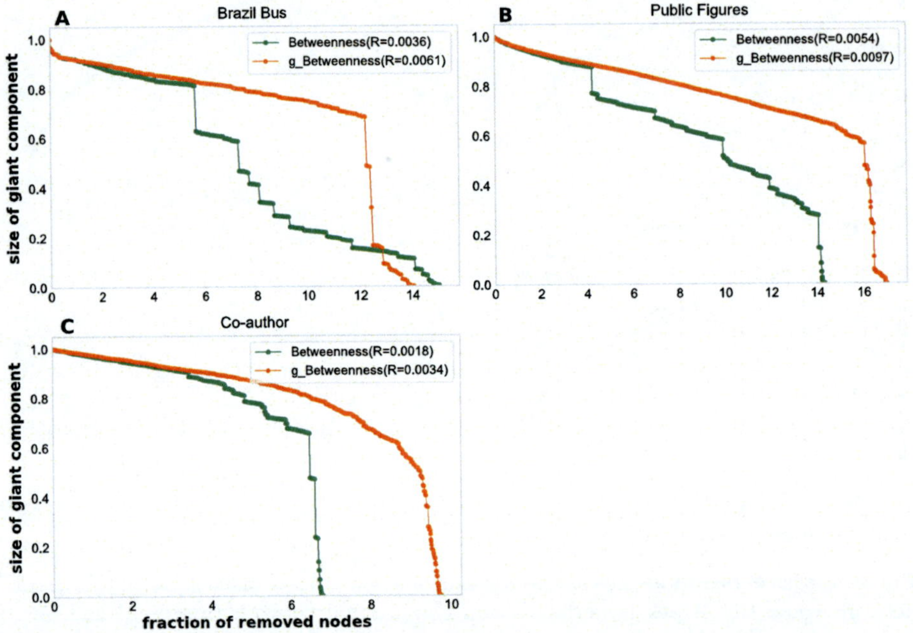

Fig. 4. Betweeness-based attacks: Evolution of the relative size of the LCC of the original network when the attack is performed on the original network (in green) and on its global components (in orange) A) Brazil Bus network B) Public Figures network C) Co-author network. The figures contain also the R values.

Betweenness Attacks. Figures 4 allows us to compare the two strategies when the attack uses the Betweenness centrality to rank the nodes. It appears that the attacks on the original networks are globally more effective according to the R Values. Note that the impact of the attacks on the entire network and the global components are comparable for a budget lower than 4% of removed nodes. Beyond this value, the LCC decreases with a sharp drop for the attacks on the original network, while it decreases linearly for the attacks on the global components. Indeed, Betweenness performs exceptionally in modular networks because the intercommunity links have high betweenness values. Therefore, it progressively disconnects the subnetworks in the original network. In contrast, it is ineffective on the global component, which does not contain subnetworks. The curves for the Bus Brazil network are slightly different. One can notice that while for Public Figures and Co-author networks, the maximum budget to dismantle the network is lower for the betweenness attacks on the original network, it is slightly higher for Bus Brazil.

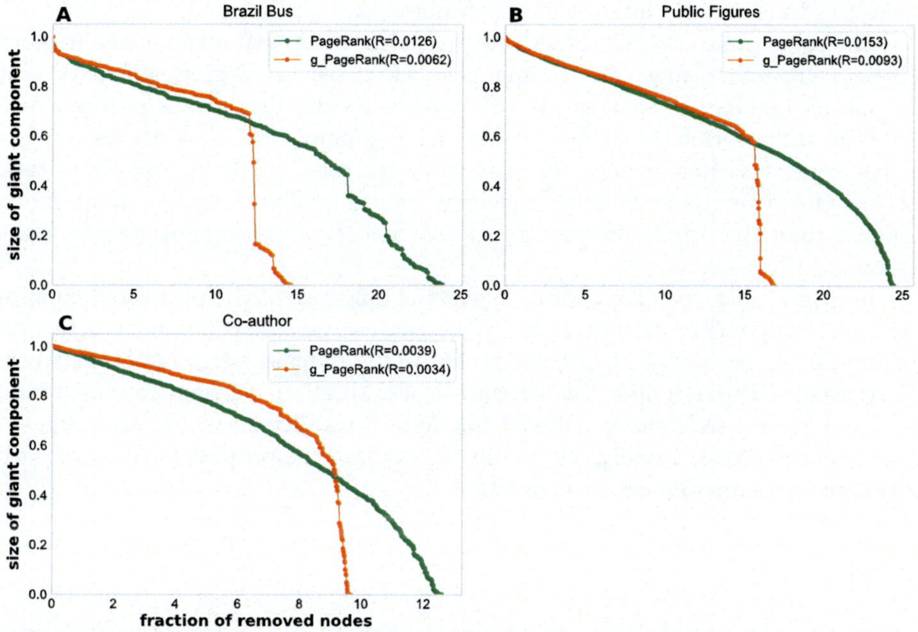

Fig. 5. PageRank-based attacks: Evolution of the relative size of the LCC of the original network when the attack is performed on the original network (in green) and on its global components (in orange) A) Brazil Bus network B) Public Figures network C) Co-author network. The figures contain also the R values.

PageRank Attack. Figures 5 reports the evolution of the LCC for the three networks for the PageRank attacks. The attacks of the global components outperform the attacks of the original networks. Indeed, the R values resulting from the attacks on the global components are smaller. Furthermore, one obtains these interesting results with a smaller budget. Nevertheless, at low and medium budget, one can see that the Brazil Bus and Co-author networks are more vulnerable to removing nodes from the original network. This observation is more pronounced in the Co-author networks. In the Public Figures network, the attacks on the original network and the global components have a similar effect until about 15.5% of the nodes are eliminated.

5 Conclusion

This paper proposes a targeted attack framework to dismantle a network. Rather than attacking the original network based on a given centrality measure, we suggest operating on the network's global components. We conduct a preliminary empirical investigation with three real-world networks of prominent centrality measures to assess the interest of this framework.

The robustness analysis shows that targeting the global components is more efficient than targeting the original network based on degree centrality and PageRank ranking. In contrast, the Betweenness centrality attack on the entire network outperforms the attack on the global components. Nevertheless, differences are not so pronounced. In that case, the main advantage of the attack on the global components is its efficiency. Indeed, global components are much smaller than the original networks, so that betweenness computation is much faster.

In future work, we plan to investigate the influence of the uncovered component structure on the results. Indeed, one can use various community detection or multi-core periphery algorithms to extract the dense parts of the network. Furthermore, through an extended empirical evaluation, we want to understand better the relations between network topology, centrality measures, and targeted attack effectiveness. Finally, we intend to compare the proposed approach with alternative community-aware attacks.

References

1. Wandelt, S., Sun, X., Feng, D., Zanin, M., Havlin, S.: A comparative analysis of approaches to network-dismantling. Sci. Rep. Ser. **8**, 1 (2018)
2. Qian, B., Zhang, N.: Topology and robustness of weighted air transport networks in multi-airport region. Sustainability **14**, 6832 (2022)
3. Sun, X., Gollnick, V., Wandelt, S.: Robustness analysis metrics for worldwide airport network: a comprehensive study. Chin. J. Aeronaut. Ser. **30**, 500 (2017)
4. Wu, Z., Braunstein, L.A., Colizza, V., Cohen, R., Havlin, S., Stanley, H.E.: Optimal paths in complex networks with correlated weights: the worldwide airport network. Phys. Rev. E Ser. **74**, 056104 (2006)
5. Wandelt, S., Shi, X., Sun, X., Zanin, M.: Community detection boosts network dismantling on real-world networks. IEEE Access Ser. **8**, 111954 (2020)
6. Musciotto, F., Micciché, S.: Exploring the landscape of community-based dismantling strategies, arXiv preprint arXiv:2209.14077 (2022)
7. Diop, I.M., Cherifi, C., Diallo, C., Cherifi, H.: Revealing the component structure of the world air transportation network. Appl. Netw. Sci. **6**(1), 1–50 (2021). https://doi.org/10.1007/s41109-021-00430-2

8. Diop, I.M., Diallo, C., Cherifi, C., Cherifi, H., et al.: Robustness analysis of the regional and interregional components of the weighted world air transportation network. Complexity **2022**, 6595314 (2022)
9. Diop, I.M., Diallo, C., Cherifi, C., Cherifi, H.: Targeted attacks on the world air transportation network: impact on its regional structure, arXiv e-prints (2022)
10. Chakraborty, D., Singh, A., Cherifi, H.: Immunization strategies based on the overlapping nodes in networks with community structure. In: Nguyen, H.T.T., Snasel, V. (eds.) CSoNet 2016. LNCS, vol. 9795, pp. 62–73. Springer, Cham (2016). https://doi.org/10.1007/978-3-319-42345-6_6
11. Gupta, N., Singh, A., Cherifi, H.: Community-based immunization strategies for epidemic control, in 2015 7th international conference on Communication Systems and Networks (COMSNETS), pp. 1–6, IEEE (2015)
12. Kumar, M., Singh, A., Cherifi, H.: An efficient immunization strategy using overlapping nodes and its neighborhoods. Companion Proc. Web Conf. **2018**, 1269–1275 (2018)
13. Ibnoulouafi, A., El Haziti, M., Cherifi, H.: M-centrality: identifying key nodes based on global position and local degree variation. J. Statist. Mech. Theory Exp. Ser. **2018**, 073407 (2018)
14. Rajeh, S., Savonnet, M., Leclercq, E., Cherifi, H.: Interplay between hierarchy and centrality in complex networks. IEEE Access Ser. **8**, 129717 (2020)
15. Lü, L., Chen, D., Ren, X.-L., Zhang, Q.-M., Zhang, Y.-C., Zhou, T.: Vital nodes identification in complex networks. Phys. Rep. Ser. **650**, 1 (2016)
16. Schneider, C.M., Moreira, A.A., Andrade, J.S., Jr., Havlin, S., Herrmann, H.J.: Mitigation of malicious attacks on networks. Proc. Natl. Acad. Sci. Ser. **108**, 3838 (2011)
17. Alves, L.G., Aleta, A., Rodrigues, F.A., Moreno, Y., Amaral, L.A.N.: Centrality anomalies in complex networks as a result of model over-simplification. New J. Phys. Ser. **22**, 013043 (2020)
18. Rozemberczki, B., Davies, R., Sarkar, R., Sutton, C.: GemSec: graph embedding with self clustering, in Proceedings of the 2019 IEEE/ACM International Conference on Advances in Social Networks Analysis and Mining, pp. 65–72 (2019)
19. Newman, M.E.: The structure of scientific collaboration networks. Proc. Natl. Acad. Sci. Ser. **98**, 404 (2001)

The Effect of Link Recommendation Algorithms on Network Centrality Disparities

Timo Debono(✉) and Fernando P. Santos(✉)

Informatics Institute, University of Amsterdam, Science Park 900, 1098XH Amsterdam, The Netherlands
timo.debono@transferwise.com, f.p.santos@uva.nl

Abstract. Link recommendation algorithms are nowadays ubiquitous in online social networks. It is fundamental to understand their impact in users' capacity to spread information and the network centrality of different groups. This paper investigates the effect of the Stochastic Approach for Link-Structure Analysis (SALSA), a popular link recommendation algorithm, on network centrality and disparate exposure in networks composed of a majority and minority group. While previous works mainly focus on exposure inequalities along degree centrality (i.e., how frequently individuals from different groups are recommended), it remains unknown how link recommendation impacts groups' visibility along different network centrality measures. We use two different centrality metrics that capture more granular information: betweenness (BC) and closeness (CC) centrality. Resorting to synthetic networks and as in previous works, we observe that SALSA can reduce the relative degree centrality of a minority group in a network, when both groups are homophilic. A similar conclusion is observed for BC. Conversely, we observe that CC of both minority and majority groups increases after SALSA recommendations are followed. Lastly, we propose the concept of *kn-Interventions* to mitigate disparate exposure effects and demonstrate that this simple intervention has a positive effect on the minority's visibility. Our findings elucidate the need to consider different measures capturing minorities' visibility when evaluating the effects of link recommendation in social contexts.

Keywords: Link recommendation · Algorithmic bias · Network centrality

1 Introduction

Link recommendation algorithms, i.e., algorithms that suggest new connections to users in a network, are a central component of current social media platforms [6,8,10,12,25]. These algorithms can nudge new connections and directly impact the growth dynamics of a network, as well as the visibility of different groups and their capacity to spread information widely. While typical metrics to evaluate the success of link recommendation algorithms include the fraction of recommendations followed by people, we must also consider the potential impacts that such algorithms have on social dynamics and groups' visibility.

A. S. Teixeira et al. (Eds.): CompleNet 2023, SPCOM, pp. 74–85, 2023.
https://doi.org/10.1007/978-3-031-28276-8_7

Prior works, e.g. [10,24,27], suggest that link recommendation systems do affect group-level visibility. Such effects are noted to depend on the level of homophily, i.e., the extent to which groups establish in-group connections as opposed to out-group connections. These works also imply that in-degree is an indication of visibility in a network. The visibility of groups extends beyond degree centrality, however, especially in social contexts [3,5]. The goal of this research is to investigate the effects of link recommendation algorithms on group visibility by using different factors that go beyond node degree and capture more granular facets of influence, specifically *betweenness* (BC) and *closeness* (CC) centrality. Here we study the impact of link recommendation algorithms on such centrality metrics and propose an intervention to mitigate disparate effects.

We focus on one main question: *What is the impact of link recommendation on groups' network centrality?* and divide our analysis in two sub-questions:

- RQ1: How does link recommendation affect BC and CC of different groups?
- RQ2: What interventions can be designed to mitigate disparate visibility effects arising as a result of the link recommendation process?

By answering these questions, this work adds new insights to the growing research on fairness in link recommendation systems [10–12,24,25]. First, we perform a simulation-based study on the effect of the Stochastic Approach for Link-Structure Analysis (SALSA) on group visibility and network centrality, taking into account centrality measures other than node degree (BC and CC). Second, we propose the concept of *kn-Interventions* to mitigate potential disparate visibility effects as a result of the algorithm's recommendations.

The remainder of this paper is composed of five sections: Following an overview of related research, we delineate the various methodologies applied throughout this work and describe the experimental setup, as well as the data we used. We follow with the presentation of the results and a discussion thereof, including identified limitations and future research directions. Lastly, we conclude with a summary of this work.

2 Related Work

2.1 Effects of Link Recommendation Systems on Fairness and Long-term Network Dynamics

Emphasis has been placed on developing accurate link recommendation algorithms whose suggestions individuals are likely to follow [1,7,13,31]. However, it is also fundamental to understand how such algorithms impact information and opinions' spread in networks [25], and how, in turn, different groups are affected in their visibility and capacity to spread and receive information.

Fabbri et al. conclude that the impact of link recommendation systems on groups' visibility is a function of homophily. By homophily we mean preferentially connecting with similar individuals, e.g., members of the same group. Regardless of the minority's level of homophily, among nodes with similar in-degree, those belonging to the majority are more likely to be recommended [10].

In a follow-up study, Fabbri et al. simulate the long-term consequences of repeatedly adding recommended links into a bi-populated network. They show that a minority group can benefit disproportionately from link recommendations if it is homophilic enough. Conversely, this effect is reversed in the case of a heterophilic minority [11]. Differing from our work, [10, 11] explicitly use in-degree to assess node popularity and visibility.

Espín-Noboa et al. study the effect of recommendations generated by PageRank and Who-to-follow (WTF) along two dimensions, inequality and inequity. In this context, inequality refers to the distribution of importance among individual nodes in a network, whereas inequity refers to group-level fairness. They measure inequality by analyzing the skewness of the rank distribution of nodes that the two algorithms suggest in terms of the Gini coefficient. To evaluate inequity, they investigate how well-represented the minority group is among the recommended nodes, relative to the proportion of the network's minority size. Their findings suggest that the two studied algorithms reduce, replicate, and amplify minority representation among top-k recommendations when the majority is homophilic, neutral, and heterophilic, respectively [24]. Similar to [10] and [11], they use the in-degree distribution of top ranks to assess inequality and inequity.

Ferrara et al. analyze how different recommendation algorithms, including PageRank, affect network structure and minority visibility in bi-populated directed networks over time. They find that most algorithms tend to recommend nodes with high in-degree, contributing to inequality as measured by the Gini coefficient of the in-degree distribution. Moreover, they study the effect of homophily and find that when both minority and majority are heterophilic, the visibility of minorities is reduced. They define the visibility of the minority group as the fraction of nodes out of the top-10% recommended nodes that belong to the minority group [12]. They also use some notion of degree to assess the effect of recommendation systems on network structure and minority exposure.

Santos et al. study the impact of link recommendation algorithms, which tend to recommend links based on the number of common connections, on opinion polarization and echo chambers. They find that preferentially establishing links with structurally similar nodes (i.e., sharing many neighbors) results in networks composed of independent modules that are amenable to opinion polarization [25].

Along similar lines, Cinus et al. investigate the effect of link recommenders, including SALSA, on echo chambers and polarization in an opinion dynamics setting. They find that such algorithms can lead to an increase in polarization, given the a priori presence of homophily in a network. In cases where echo chambers are already present in the network, the effect of recommenders is negligible. Moreover, they evaluate the impact of intervention strategies that act on top of the recommendation system to mitigate the formation of echo chambers: given a recommended edge, they replace it with probability p with another edge, according to one of three intervention strategies (random uniform, opinion diversity, and degree-based). They conclude that the opinion diversity strategy is most effective [8]. This idea of designing interventions to mitigate disparities motivates the approach we introduce in Sect. 3.4.

2.2 Node Influence Quantification

The position of a node in a network determines its power [5,9], status [16], and ability to receive and spread information [5,19,26], among others [29]. Understanding how information diffuses can thereby benefit from identifying the centrality properties of the nodes that initially seed information in a network. In this context, measures of betweenness were proposed to capture a node's control over the flow of information in a network [23]. The ability of a node to control information flow constitutes its influence in the network [19,26].

The concepts of power, status, and information control in a network tie back to the idea of capturing a node's influence. Measuring the influence of a node is commonly abstracted by some notion of centrality [23]. In general, centrality measures allow to rank nodes according to their structural significance and refer to a node's ability to affect global network processes, e.g., information diffusion in social networks [17] or the spreading of diseases within a population [2].

One of the most fundamental notions of centrality is node degree, i.e., the number of adjacent nodes. Despite its widespread adoption in settings of information diffusion, influence and visibility [10–12,24], several studies demonstrate that network structure features alone, including node degree, might not provide a full picture of individuals' influence [3,5,9,30]. More recent approaches, e.g., [21], seek to integrate structural properties and node characteristics.

3 Methodology

3.1 Synthetic Network Data

In this study, we consider directed networks with attributes. A directed network is defined as $G = (V, E, C)$, where V is the set of nodes, E is the set of directed edges, and $C \rightarrow V : \{0, 1\}$ is a function mapping each node $v_i \in V$ to a binary label representing group affiliation [23]. In our work, C splits the nodes into two distinct groups, a majority M (0) and a minority m (1). For simplicity, we exclusively consider binary labels, but this mapping can generally be extended to multiple labels.

We generate one such synthetic network based on the implementation proposed in [24], using the Directed network with Preferential Attachment and Homophily (DPAH) model, where both groups are highly homophilic ($h_m = h_M = 0.8$), i.e., nodes preferably connect with nodes from the same group. This network consists of $|V| = 10,000$ nodes, with minority fraction $f_m = 0.2$, and $|E| = 100,000$ edges. The power-law parameters of the activity distributions are set to $\gamma_m = \gamma_M = 2.5$.

Our choice to focus on this specific configuration is motivated by the fact that this scenario most closely resembles a real-world setting: individuals belonging to the same group are more likely to interact, i.e., establish a link with each other [22].

3.2 Centrality Measures

As expanded upon in Sect. 2.2, quantifying a node's influence in a network has multiple facets, and evaluating in-degree, despite commonly used, is not per se sufficient in social contexts. To evaluate the relative influence of the minority and majority groups and their ability to control information, we make use of different centrality measures to more accurately reflect this. We further abstract these two facets under the terms exposure or visibility. In this work, we consider the measures of degree centrality (DC), BC, and CC.

The degree of a node is defined as the number of connections a node has. We study the in-degree of a node, i.e., the number of incoming edges [18,29].

Betweenness is a measure of node centrality based on shortest paths between any two nodes in a network. Precisely, it captures the degree of influence a node v_u has over information flow in a network by measuring the share of shortest paths between two nodes v_i and v_j that pass through v_u, relative to the total number of shortest paths between v_i and v_j, for all nodes v_i,$v_j \in V$ [18].

Closeness assumes that nodes that are closer to other nodes in the network can spread information more efficiently throughout the network [4].

3.3 Experimental Setup

Given the network generated as per Sect. 3.1, we first calculate the centrality measures described in Sect. 3.2 for every node $v_i \in V$ and then aggregate the results for minority and majority nodes to obtain group-level averages. Subsequently, we draw a random node sample without replacement from V. Let V^* be this random sample.

To simulate a network's recommendation dynamics, we adapt the Stochastic Approach for Link-Structure Analysis (SALSA), a popular recommendation algorithm originally introduced in [20], based on the implementation in [24]. For every node v_i^* in V^*, we use SALSA to suggest new connections. Precisely, we proceed as outlined in Algorithm 1.

In our framework, k is the number of nodes added, per node, out of the list of ranked recommendations generated by SALSA and p is the sample fraction. During the recommendation process, we set $p = 0.2$, i.e., we consider 20% of the network's nodes, which equals 2,000 nodes in total. This choice is motivated by the idea that in a real-world network, not every node is active at any given time. Furthermore, we set $k = 5$ and define the edges that are added for every node under consideration to be the top-k of the recommended links. Given that the networks we consider in this work have 100,000 edges, we set $k = 5$ to consider up to 10,000 edges (i.e., 10%) during the recommendation process.

As noted in [8,11,12], the role of the acceptance policy in altering the network topology, and eventually the simulation's results, is negligible.

In addition, for every new edge that is established, we first remove an existing out-edge. We do so to keep the edge density approximately constant and ensure that any results we obtain are a result of the algorithm's recommendations and not of an increase in a node's degree. Moreover, we remove out-edges before

Algorithm 1. Recommendation process

Input: $G = (V, E, C)$, k, p
Output: $G^* = (V, E^*, C)$

$sample\_size \leftarrow |V| * p$ ▷ determine sample size
$V^* \leftarrow random\_sample(V, sample\_size)$ ▷ randomly sample from all nodes
$recommendations \leftarrow SALSA(G)$ ▷ apply SALSA to obtain recommendations
for v_i^* in V^* **do**
    $out\_edges_i \leftarrow out\_edges(G, v_i^*)$ ▷ obtain out-edges
    $k \leftarrow min(len(out\_edges\_i), k)$ ▷ update k to keep edge density constant
    $delete\_random\_edges(G, out\_edges_i, k)$ ▷ delete k random out-edges
    $recommendations_i \leftarrow recommendations[v_i^*]$ ▷ retrieve node recommendations
    for v_j in $recommendations_i[: k]$ **do** ▷ add top-k recommended nodes
        $add\_edge(G, (v_i^*, v_j))$
    end for
end for

adding recommended links. The number of deleted out-edges is determined by the minimum between a node's existing out-edges and the number of edges to be added. We make this design choice to ensure that any measurable changes are a direct result of the algorithm's recommendations. If we were to first add the recommendations and then randomly delete as many out-edges for each node, we might unwantedly delete the added links and therefore at least partially neglect the algorithm's effect.

Following the application of Algorithm 1, we recalculate the centrality measures for G^*, i.e., network G, but with an updated set of edges, E^*, and again aggregate the results to obtain group-level averages. Subsequently, for each of the two groups, we record the relative metric changes between the initial values and the values after the recalculation. We repeat this procedure $z = 10$ times and report the final results as the metric changes averaged over z runs, together with the corresponding standard deviations.

To evaluate these results using the metrics described in Sect. 3.2, we consider either of the two groups to gain exposure in cases where the metric change for one group is greater than the corresponding change for the other group. This definition also includes scenarios where both groups experience a decrease as a result of the algorithm's recommendations, but one of the two groups suffers a relatively smaller decline.

3.4 kn-Interventions

Fundamentally, the goal of any intervention policy in this context is the mitigation of disparate exposure effects induced by link recommendation dynamics.

In this section, we introduce the concept of *kn-Interventions.* We define such an intervention as $I_{k,n}$, where k is the number of nodes to be added out of the ranked list of recommendations produced by the recommendation algorithm (cf. Algorithm 1) for each node, R, and n is the minimum number of nodes of the opposing group that must be part of the set of recommended nodes.

Precisely, we proceed as follows: for any given node $v_i \in V$, we first obtain the top-k recommendations generated by the algorithm, K, with $K \subset R$, as well as the class membership of v_i, $c_i \in 0, 1$. With n determining the minimum number of nodes in K that must have class $\hat{c} = 1 - c_i$, we determine if set K meets this condition and iteratively remove the last element in K that belongs to class c_i until the constraint introduced by n is met. Putting this intervention policy into the context of the experimental setup and following a vanilla simulation that corresponds to $n = 0$, we simulate the network's dynamics with varying n. Specifically, we consider $I_{5,n}$ for $n \in \{1, 2, 3, 4\}$. Informally, with these interventions we guarantee that at least n members of the minority group are included in the recommendations; while the positive effect on the minority degree centrality is obvious, we will evaluate whether the positive effects extend to BC and CC.

4 Results

First, we show how the effect of link recommendation dynamics induced by SALSA affects the metrics introduced in Sect. 3.2 (RQ1). Second, we demonstrate how the intervention proposed in Sect. 3.4 impacts the evaluation metrics (RQ2). Every result presented here will be discussed further in Sect. 5.

4.1 How Does Link Recommendation Affect BC and CC of Groups in a Network?

Given our initially stated motivation, we are interested in comparing DC results to BC and CC results. From the results presented in Fig. 1, we find that the evaluation of the results as measured by DC leads to similar interpretations when measuring the changes in BC. That is, when one group profits from the recommendation algorithm relative to the other group as measured by DC, then evaluating the results using BC reveals the same conclusion. Regardless, we notice that signs and magnitudes of the measured changes differ between these two metrics. Generally, we record a relative disadvantage for the minority by either metric.

Evaluating the results by CC, we find that both the minority and the majority record a gain after the simulation. In fact, the minority's gain is greater than the majority's. These results suggest a different interpretation to the DC results.

4.2 What Interventions Can Be Designed to Mitigate Disparate Visibility Effects Arising as a Result of the Link Recommendation Process?

We proposed the idea of *kn-Interventions* as one possible design for mitigating visibility discrepancies between groups in online social networks. Figure 2 shows the simulation results for different n. As n increases, we observe a strictly positive increase in DC and BC for the minority and the inverse development thereof for the majority. In addition, besides the scenario where $n = 1$, the increases for the

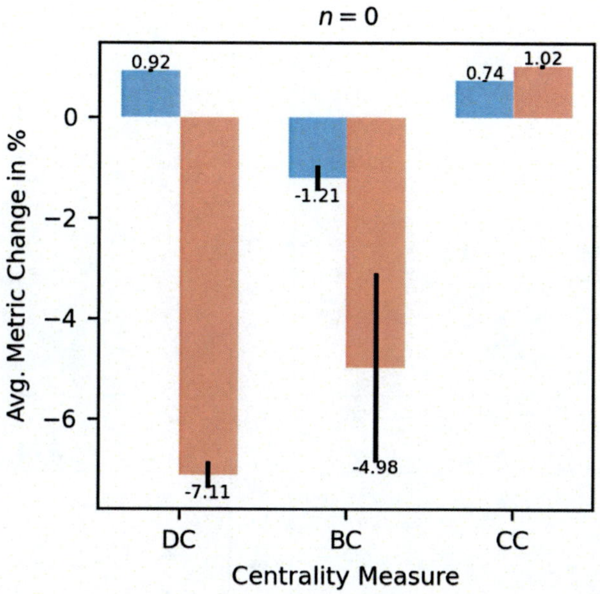

Fig. 1. Average change (in percentage) in centrality measures between the original network and the network after adding edges using SALSA. Results for the minority are in red (right-side bars) and for the majority in blue (left-side bars). The y-axis indicates the average magnitude and direction of change over $z = 10$ runs and the error bars specify the standard deviation over all runs. (Color figure online)

minority are always more pronounced when evaluated by BC. Looking at CC, however, although the minority gains more relative to the majority irrespective of n, the changes for both groups exhibit a converging behavior as n grows.

5 Discussion

Other works, [10–12, 24, 27], have used some notion of degree or ranking in terms of recommendation frequency to investigate disparate exposure effects between two groups in a network as a result of link recommendation dynamics. We extend these analyses to two other measures of network influence (BC and CC). Given the group-level homophily configuration we considered for our analysis, our results as determined by DC match those presented in [12, 24].

When evaluating these results by BC, the interpretation is comparable to that of DC, i.e., that the minority group suffers from negative exposure, relative to the majority group. This observation is, at least partially, in line with the findings presented in [28], where it is shown that DC and BC exhibit a moderate-to-high correlation. Moreover, we find that group-level changes in CC are small in magnitude, relative to changes in other metrics, and both groups record an increase in CC, with the increase for the minority group greater than

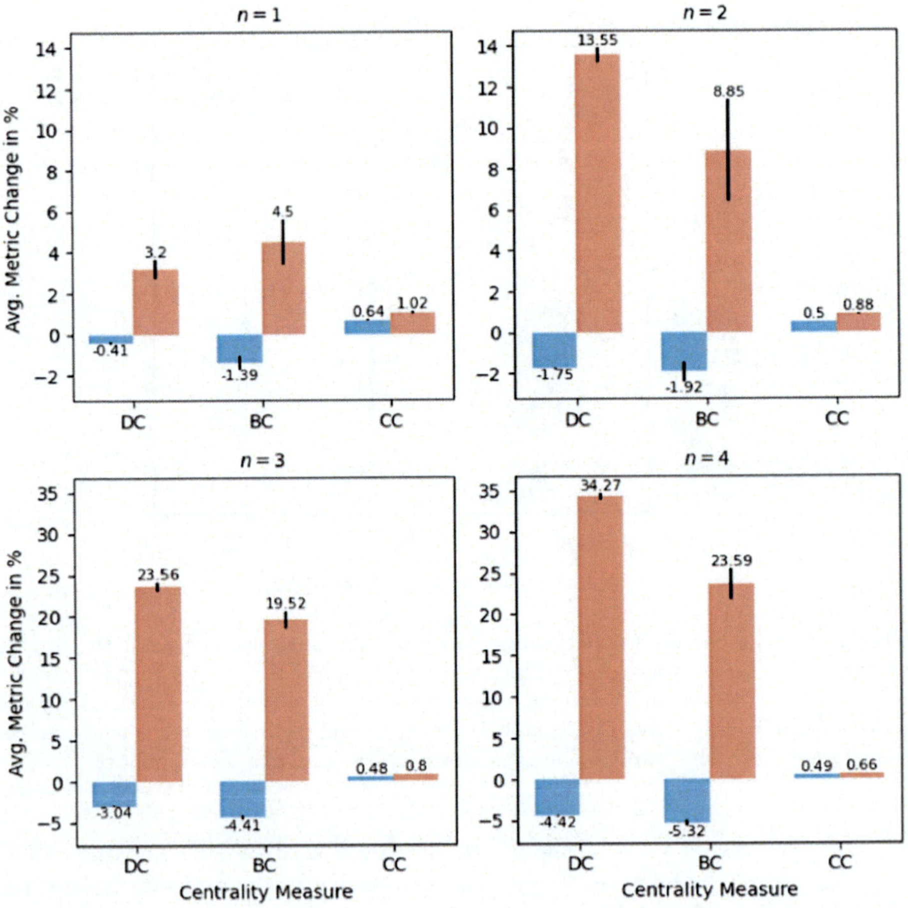

Fig. 2. Average change (in percentage) in centrality measures between the original network and the network after adding edges using SALSA under varying constraints as determined by parameter n. Results for the minority are in red (right-side bars) and for the majority in blue (left-side bars). The y-axis indicates the average magnitude and direction of change over $z = 10$ runs and the error bars specify the standard deviation over all runs. The x-axis shows the respective centrality measures. (Color figure online)

the increase for the majority group. While around 20% of the nodes we are making recommendations to belong to the minority, almost all newly established edges for those nodes will be of type $m \rightarrow M$, given that majority nodes dominate the recommendations. Therefore, minority nodes will be, on average, closer to a greater number of nodes after the recommendation. These results underline the need for more granular evaluation methods when investigating the effects of link recommendation systems on groups' visibility in social networks.

Additionally, we introduced the idea of *kn-Interventions* and evaluated the effectiveness thereof for different values of n. We imposed a constraint on how new links are established *after* recommendations had been generated. As can be seen in Fig. 2, setting $n = 1$ already alters the results distinctively as compared to the baseline scenario under $n = 0$ presented in Fig. 1. By both DC and BC, the minority now experiences positive visibility. We suspect that such a change from $n = 0$ to $n = 1$ is related to the make-up of the generated recommendations. We find that the average share of nodes within the top-10 recommendations belonging to the minority is 6.48%, meaning that minority nodes are underrepresented relative to their group size (20%). Hence, it appears reasonable that forcing at least representative parity among the $k = 5$ nodes we eventually add by setting $n = 1$ will result in a positive effect for the minority. It is not surprising then, that this effect is amplified as n increases.

6 Limitations and Future Work

Following the previous observations, we identify at least four limitations of our current approach, which provide possible lines for future work. First, subsequent studies should further test the behavior of CC as n grows, e.g., by explicitly measuring the contribution of individual links to the gains in CC for the two groups. In general, the impacts of the proposed interventions should be considered for different combinations of k and n. Second, we acknowledge that our study focused on one particular link recommendation algorithm, thus leaving open the possibility, to compare different recommendation algorithms based on alternative link-prediction methods [14,15]. Third, continuing research could apply the methods proposed and implemented in this work to a wider range of synthetic and real-world social networks. Lastly, we see room for future work for the application of different measures of network influence [30].

7 Conclusion

In this work, we analyzed the effect of one particular link recommendation algorithm, i.e., SALSA, on network centrality and visibility discrepancy between two groups in a network. We studied how centrality measures that go beyond degree, i.e., BC and CC, are affected by applying SALSA. For the network we considered and in line with previous studies, we find that the minority receives less exposure, relative to the majority, when measuring the impact on visibility by changes in DC. We find similar results when evaluating the effects using BC.

Moreover, we introduced the concept of *kn-Interventions* to study the effect of intervening during the recommendation process on changes in visibility. We find that forcing the number of nodes from the opposing group to $n = 1$ when generating recommendations has a profoundly positive effect on the minority's visibility, likely as a result of the minority's general under-representation among generated recommendations. This effect is amplified as n increases.

Ultimately, our work elucidates the need for more nuanced evaluation criteria in simulation-based studies concerned with the investigation of visibility disparity in social networks.

Acknowledgements. We acknowledge the University of Amsterdam - Master Programme Information Studies for creating the conditions to perform this research and for financially supporting this publication. We thank the CompleNet anonymous reviewers for the pertinent recommendations to improve this article.

References

1. Asaph, W., Sun, S.: Improving business pages recommendation in social network using link prediction methods. Asian J. Probab. Statist. **14**, 1–2 (2021)
2. Bajardi, P., Barrat, A., Savini, L., Colizza, V.: Optimizing surveillance for livestock disease spreading through animal movements. J. R. Soc. Interface **9**, 2814–2825 (2012)
3. Banerjee, A., Chandrasekhar, A.G., Duflo, E., Jackson, M.O.: The diffusion of microfinance. Science. **341**, 1236498 (2013). https://doi.org/10.1126/science.1236498
4. Beauchamp, M.A.: An improved index of centrality. Behav. Sci. **10**, 161–163 (1965). https://doi.org/10.1002/bs.3830100205
5. Bonacich, P.: Power and centrality: a family of measures. Am. J. Sociol. **92**, 1170–1182 (1987). https://doi.org/10.1086/228631
6. Chen, J., Geyer, W., Dugan, C., Muller, M., Guy, I.: Make new friends, but keep the old: recommending people on social networking sites. In: Proceedings of the SIGCHI Conference on Human Factors in Computing Systems, pp. 201–210. CHI 2009, Association for Computing Machinery, New York, NY, USA (2009). https://doi.org/10.1145/1518701.1518735
7. Ciesielczyk, M., Szwabe, A., Morzy, M.: On efficient link recommendation in social networks using actor-fact matrices. Sci. Program. **2015**, 450215:1-450215:9 (2015)
8. Cinus, F., Minici, M., Monti, C., Bonchi, F.: The effect of people recommenders on echo chambers and polarization. In: International Conference on Web and Social Media (2021)
9. Cook, K.S., Emerson, R.M., Gillmore, M.R., Yamagishi, T.: The distribution of power in exchange networks: theory and experimental results. Am. J. Sociol. **89**, 275–305 (1983). https://doi.org/10.1086/227866
10. Fabbri, F., Bonchi, F., Boratto, L., Castillo, C.: The effect of homophily on disparate visibility of minorities in people recommender systems. In: Proceedings of the International AAAI Conference on Web and Social Media, vol. 14(1), pp. 165–175, May 2020. https://ojs.aaai.org/index.php/ICWSM/article/view/7288
11. Fabbri, F., Croci, M.L., Bonchi, F., Castillo, C.: Exposure inequality in people recommender systems: the long-term effects. In: Proceedings of the International AAAI Conference on Web and Social Media, vol. 16 (2022)
12. Ferrara, A., Espin-Noboa, L., Karimi, F., Wagner, C.: Link recommendations: their impact on network structure and minorities. In: 14th ACM Web Science Conference 2022. ACM, June 2022. https://doi.org/10.1145/3501247.3531583
13. Ghasemian, A., Galstyan, A.G., Clauset, A.: Highly accurate link prediction in networks using stacked generalization. In: WSDM 2018: The Eleventh ACM International Conference on Web Search and Data Mining (2018)

14. Ghasemian, A., Hosseinmardi, H., Galstyan, A., Airoldi, E.M., Clauset, A.: Stacking models for nearly optimal link prediction in complex networks. Proc. Natl. Acad. Sci. **117**(38), 23393–23400 (2020)
15. Ghorbanzadeh, H., Sheikhahmadi, A., Jalili, M., Sulaimany, S.: A hybrid method of link prediction in directed graphs. Expert Syst. Appl. **165**, 113896 (2021)
16. Katz, L.: A new status index derived from sociometric analysis. Psychometrika. **18**, 39–43 (1953). https://doi.org/10.1007/BF02289026
17. Kim, H., Yoneki, E.: Influential neighbours selection for information diffusion in online social networks. In: 2012 21st International Conference on Computer Communications and Networks (ICCCN), pp. 1–7 (2012)
18. Landherr, A., Friedl, B., Heidemann, J.: A critical review of centrality measures in social networks. Bus. Inf. Syst. Eng. **2**, 371–385 (2010). https://doi.org/10.1007/s12599-010-0127-3
19. Leavitt, H.J.: Some effects of certain communication patterns on group performance. J. Abnormal Soc. Psychol. **46**, 38 (1951). https://doi.org/10.1037/h0057189
20. Lempel, R., Moran, S.: Salsa: the stochastic approach for link-structure analysis. ACM Trans. Inf. Syst. **19**, 131–160 (2001)
21. Leng, Y., Sella, Y., Ruiz, R., Pentland, A.S.: Contextual centrality: going beyond network structure. Sci. Rep. **10**, 9401 (2020)
22. McPherson, M., Smith-Lovin, L., Cook, J.M.: Birds of a feather: homophily in social networks. Ann. Rev. Sociol. **27**, 415–444 (2001). https://doi.org/10.1146/annurev.soc.27.1.415
23. Newman, M.: Networks (2nd Edn.). Oxford University Press, Oxford (2018)
24. Noboa, L.E., Wagner, C., Strohmaier, M., Karimi, F.: Inequality and inequity in network-based ranking and recommendation algorithms. Sci. Rep. **12**, 2012 (2021)
25. Santos, F.P., Lelkes, Y., Levin, S.A.: Link recommendation algorithms and dynamics of polarization in online social networks. Proc. Natl. Acad. Sci. **118**(50), e2102141118 (2021)
26. Stephenson, K., Zelen, M.: Rethinking centrality: methods and examples. Soc. Netw. **11**, 1–37 (1989). https://doi.org/10.1016/0378-8733(89)90016-6
27. Stoica, A.A., Riederer, C.J., Chaintreau, A.: Algorithmic glass ceiling in social networks: the effects of social recommendations on network diversity. In: Proceedings of the 2018 World Wide Web Conference (2018)
28. Valente, T.W., Coronges, K., Lakon, C.M., Costenbader, E.: How correlated are network centrality measures? Connections **28**(1), 16–26 (2008)
29. Wan, Z., Mahajan, Y., Kang, B.W., Moore, T.J., Cho, J.H.: A survey on centrality metrics and their network resilience analysis. IEEE Access **9**, 104773–104819 (2021). https://doi.org/10.1109/access.2021.3094196
30. Zareie, A., Sheikhahmadi, A., Jalili, M., Fasaei, M.S.K.: Finding influential nodes in social networks based on neighborhood correlation coefficient. Knowl. Based Syst. **194**, 105580 (2020)
31. Zhou, L., Cui, X., Zeng, A., Fan, Y., Di, Z.: Improving diffusion-based recommendation in online rating systems. Int. J. Mod. Phys. C. **32**, 2150094 (2021)

CoreGDM: Geometric Deep Learning Network Decycling and Dismantling

Marco Grassia and Giuseppe Mangioni(✉)

Dipartimento di Ingegneria Elettrica,
Elettronica e Informatica (DIEEI) - Università degli Studi di Catania, Catania, Italy
{marco.grassia,giuseppe.mangioni}@unict.it

Abstract. Network dismantling deals with the removal of nodes or edges to disrupt the largest connected component of a network. In this work we introduce CoreGDM, a trainable algorithm for network dismantling via node-removal. The approach is based on Geometric Deep Learning and that merges the Graph Dismantling Machine (GDM) [19] framework with the CoreHD [40] algorithm, by attacking the 2-core of the network using a learnable score function in place of the degree-based one. Extensive experiments on fifteen real-world networks show that CoreGDM outperforms the original GDM formulation and the other state-of-the-art algorithms, while also being more computationally efficient.

Keywords: Network dismantling · Site percolation · Supervised learning · Geometric deep learning · Machine learning

1 Introduction and Related Works

In general, the function performed by a network (thus, by the underlying system) strongly depends on its structure. A change in topology can have, in fact, catastrophic impacts on networks and this crucial for all those networks that represent critical infrastructures, such as power-grids, natural gas delivery or transportation networks, just to mention a few. In this sense, investigating the robustness to external perturbations of a network, and in particular of targeted attacks to their units, attracted increasing attention for the implications and for policymaking, for instance to identify the vulnerabilities in order to enhance the robustness [8,9] of a critical infrastructure by increasing its resistance to faults or intentional attacks. Network Dismantling is the problem of finding the minimal set of units to attack to dismantle a network, is a computationally challenging (NP-hard) problem that is commonly approached by using heuristics, as detailed in the following. The first approaches were based on the removal of nodes according to their centrality metrics, like the degree [2,12], the intuition being that the most connected nodes would cause more damage to the network, or betweenness centrality [26], as it would break the network into clusters by removing the bridging nodes. The centrality metrics could be computed offline, before the attack, or

A. S. Teixeira et al. (Eds.): CompleNet 2023, SPCOM, pp. 86–94, 2023.
https://doi.org/10.1007/978-3-031-28276-8_8

updated online during the attack. We refer to the first class as "static" and to the second as "dynamic" attacks: while the latter usually perform better, they are more computationally expensive, since they require computing the metric many times as the network is attacked. Later, influence maximization centrality metrics like Collective Influence have also been employed [31]. Recent works also propose ad-hoc algorithms, like the Min-Sum (MS) [6], CoreHD [40] and Generalized Network Dismantling (GND) [33]. Cutting-edge approaches are based on Machine Learning, thanks to the significant improvements made in the field of Geometric Deep Learning and to Graph Neural Networks (GNNs) [23], trainable message-passing algorithms that not only leverage the node features (like classic Deep Learning algorithms) but also the network topology, and that often rooted to the Convolutional Neural Networks for the Euclidean data like images [23]. The key idea of Graph Neural Networks is to implement the convolutional operator by using the network structure to propagate information from node to node, and to learn the weights of the propagation function from the data. In fact, Graph Neural Networks have shown remarkable capabilities in many graph-related tasks, including node classification [27], link prediction [20,34], graph generation [29], or even on computationally hard problems like the Link Building [10] and the Maximum Clique Enumeration (MCE) [3] problems. Network Dismantling algorithms that involve Graph Neural Networks are GDM [19] and FINDER [17], the main difference being the training method: GDM is trained with a supervised learning approach on small synthetic networks dismantled optimally, while FINDER is trained, again on synthetic networks, but in a reinforcement learning fashion, thus the agent has to learn how to dismantle the network by itself.

The paper is organized as in the following: in Sect. 2 we introduce the proposed algorithm, in Sect. 3 we present the dataset and the experimental results, and in Sect. 4 we draw the conclusions.

2 Proposed Method

In this work, we introduce CoreGDM, a novel machine learning approach to the Network Dismantling problem. It represents a step forward the GDM framework presented in [19], as it combines the GDM framework with the CoreHD decycling algorithm [40]. Specifically, the key idea is to decycle the network by dismantling its 2-core via a learned strategy—whereas CoreHD recursively removes the nodes with the highest degree—and then, once the network is broken into a forest, apply the tree-breaking algorithm proposed in [6]. At this point, we employ the same greedy reinsertion algorithm to reinsert the nodes that are not actually needed to break the network in small components. As in the original GDM framework, the dismantling strategy is learned in a supervised manner, by training a Geometric Deep Learning model on a dataset of small synthetic networks dismantled optimally via brute-force. In particular, the model is trained to predict which nodes belong to the optimal dismantling set. The CoreGDM flow is summarized in Algorithm 1.

The advantages of such combination are twofold. On one hand, the strategies learned by GDM have been proven to offer a better performance than the state-of-the-art heuristics, and specifically better than the ones based on the node degree centrality employed by CoreHD. On the other hand, attacking the Largest Connected Component of the 2-core of (sparse) networks is faster than attacking the whole network, as the number of nodes and edges to be processed is reduced, which results in a significant speed-up. In fact, while the computational complexity is untouched, the number of nodes $|V_{\text{LCC}}^{2-\text{core}}|$ and edges $|E_{\text{LCC}}^{2-\text{core}}|$ in the Largest Connected Component of the 2-core is expected do be smaller, especially in sparse networks, thus the number of message-passing iterations performed by the Graph Neural Networks is reduced.

Algorithm 1. CoreGDM algorithm with static GDM predictions

```
Input: A network G = (V, E), a dismantling threshold θ
Compute the node features
Predict with GDM the removal scores (p_i) of nodes in G^2, the 2-core of G
while G^2 = (V^{2-core}, E^{2-core}), is not empty do
    Compute G^2_LCC, the Largest Connected Component of G^2
    Get x, the node with the highest predicted removal score p_i in G^2_LCC
    Remove x from G^2_LCC
    if The Largest Connected Component of G is smaller than θ then stop iterating
    end if
end while
Break the forest G using the Tree-breaker algorithm [6]
Reintroduce the nodes greedily using the Reinsertion Algorithm [6], until θ is
  reached
```

3 Dataset and Results

We test our approach on a set of fifteen real-world networks that represent systems from various domains (i.e., biological, infrastructural, social, and technological). The full list of the networks used for evaluation is available in Table 1, which reports the name, the category and the references. Moreover, in Table 2 we report a set of topological measures. In particular, we provide: the number of nodes $|V|$ and of edges $|E|$; the density, which measures the sparsity of the network as it is the ratio between the number of existing links over the number of all possible links; the average degree, which is the average number of links per node, and the degree assortativity, which measures the degree correlation between nodes; the average local clustering coefficient, which measures the fraction of possible triangles that exist through a given node; the transitivity, which is the fraction of all possible triangles; and the maximum and average K-core number. The networks exhibit a wide range of topological properties, including different sizes, densities, degree assortativity values and K-core. Regarding the training data and methodology, we use the same settings as GDM [19]. More in detail, we use Geometric Deep Learning models with Graph Attention Network (GAT) [37] layers and a Multi-Layer Perceptron for regression, trained in a supervised fashion on a set of small synthetic networks with 25 nodes each. The train networks are generated using the Barabási-Albert (BA), the Erdős-Rényi (ER) and the Static Power law generational models that are implemented in *igraph* [14] and *NetworkX* [22]; each synthetic network is dismantled optimally to shrink the Largest Connected Component to a target size of $\sim$ 18% via a brute-force attack, and nodes are assigned a training label that depends on their presence in the optimal dismantling set(s). For more details, we refer the Reader to the original paper [19].

We compare our approach against the best performing state-of-the-art algorithms that include a reinsertion phase, and also against the original GDM proposal. The performance of the algorithms are reported in terms of area under the dismantling curve (AUC), a metric that captures the quality of the entire dismantling process as it accounts for the number of nodes in the Largest Connected Component (LCC) at each removal step, and that has also the advantage of allowing comparisons over large datasets. We note that we set the dismantling target to $\sim$ 10% of the network size. In detail, for each heuristic and network, we calculate the AUC value using Simpson's rule on the $y(r) = |V_{\mathrm{LCC}}(r)|/|V|$ curve, where $|V_{\mathrm{LCC}}(r)|$ is the size of the Largest Connected Component as a function of the number of removed nodes r; the lower the AUC, the better on average during the dismantling process. For sake of simplicity, we report the relative values of the AUC, meaning that we scale (and then multiply by 100) the AUC value to the one of CoreGDM for the same network: a value lower than 100 means that the algorithm outperforms CoreGDM, and vice-versa. The results, shown in Table 3, show that the static version of our approach outperforms (on average) all the other algorithms, including the original GDM proposal. In particular, while there is no overall best performer, as some methods provide better performance in specific networks, the mean AUC of CoreGDM is lower than GDM's (the second-best performer) about 2.5%, the AUC values range from 95.5% to 120.1%. Other algorithms show even higher mean AUC values, like GND with 109.1% or CoreHD with 122.2%. Moreover, CoreGDM outperforms all the other algorithms in many networks like the two powergrids, librec-filmtrust-trust, maayan-figeys, moreno_crime_projected, and moreno_train, often with a large margin.

The code is implemented in Python on top of PyTorch, uses PyTorch Geometric (PyG) [18] for the GNN layers. The feature computation and the dismantling process is implemented in graph-tool [32].

Table 1. Real-world test networks. The networks used to evaluate our approach. For each network, we report the name, the category it belongs to and the references.

Network	Name	Category	References
ARK201012_LCC	CAIDA ARK (Dec 2010) (LCC)	Infrastructure	[7]
eu-powergrid	SciGRID Power Europe	Power	[30]
internet-topology	Internet (AS) topology	Infrastructure	[41]
librec-filmtrust-trust	FilmTrust trust network	Social	[21]
maayan-figeys	Human protein (Figeys)	Metabolic	[16]
moreno_crime_projected	Crime (projection)	Social	[1]
moreno_propro	Protein	Metabolic	[13, 24, 36]
moreno_train	Train bombing terrorist contacts	Human contact	[25]
munmun_digg_reply_LCC	Digg social network replies (LCC)	Communication	[11]
opsahl-powergrid	US power grid	Infrastructure	[39]
power-eris1176	Power network problem	Power	[35]
slashdot-zoo	Slashdot Zoo	Social	[28]
subelj_jung-j	JUNG and Javax dependency network	Software	[38]
web-EPA	Pages linking to epa.gov	Hyperlink	[15]
web-webbase-2001	Web network	Hyperlink	[4, 5]

Table 2. Topological measures of the test networks. $|V|$ and $|E|$ are the number of nodes and edges in the network, "T" stands for Transitivity.

Network				Degree		Avg.		K-core num.	
	$\|V\|$	$\|E\|$	Density	avg.	assort.	LCC	T	max.	avg.
ARK201012	29K	78K	1,81E-04	5,32	−0,04	0,38	0,02	33	2, 71
eu-powergrid	1K	2K	1,69E-03	2,48	−0,08	0,13	0,12	2	1, 58
internet-topology	35K	108K	1,78E-04	6,20	−0,04	0,29	0,05	63	3, 21
librec-filmtrust-trust	874	1K	3,43E-03	3,00	0,10	0,16	0,19	8	1, 79
maayan-figeys	2K	6K	2,57E-03	5,75	−0,14	0,04	0,01	10	3, 02
moreno_crime_projected	754	2K	7,49E-03	5,64	0,15	0,73	0,58	17	4, 62
moreno_propro	2K	2K	1,30E-03	2,44	−0,11	0,07	0,06	5	1, 48
moreno_train	64	243	1,21E-01	7,59	0,07	0,62	0,56	10	5, 61
munmun_digg_reply	30K	85K	1,93E-04	5,72	−0,01	0,01	0,01	9	2, 96
opsahl-powergrid	5K	7K	5,40E-04	2,67	−0,07	0,08	0,10	5	1, 74
power-eris1176	1K	10K	1,43E-02	16,78	0,96	0,43	0,94	81	15, 68
slashdot-zoo	79K	468K	1,49E-04	11,82	−0,03	0,06	0,02	54	6, 04
subelj_jung-j	6K	50K	2,69E-03	16,43	−0,13	0,68	0,01	65	9, 11
web-EPA	4K	9K	9,77E-04	4,17	−0,25	0,07	0,01	6	2, 24
web-webbase-2001	16K	26K	1,98E-04	3,19	−0,05	0,22	0,02	32	1, 90

Table 3. Results on the real-world test networks. Per-method area under the curve (AUC) of real-world networks dismantling. The lower, the better. The dismantling target for each method is 10% of the network size. We compute the AUC value by integrating the $LCC(x)/|N|$ values using Simpson's rule, the resulting values are scaled to the one of our approach (CoreGDM) for the same network. "+R" means that the reinsertion phase is performed. CoreHD, Collective Influence (CI) and FINDER are compared to other +R algorithms as they include the reinsertion phase. We also note that we test Collective Influence with three different ball radius ℓ values (1, 2, 3).

Heuristic network	CoreGDM	GDM +R	GND +R	CoreHD	MS +R	CI $\ell-2$	CI $\ell-3$	CI $\ell-1$	FINDER
ARK201012_LCC	100.0	101.0	93.6	98.9	102.4	122.4	128.3	120.9	111.8
eu-powergrid	100.0	103.3	106.7	133.8	148.8	175.5	147.9	173.1	237.6
internet-topology	100.0	101.6	90.8	107.3	109.0	110.4	116.7	107.2	106.4
librec-filmtrust-trust	100.0	100.2	106.7	119.4	110.2	109.5	109.4	114.9	126.7
maayan-figeys	100.0	100.4	102.6	126.7	126.4	102.0	110.0	108.0	101.8
moreno_crime_projected	100.0	101.1	108.5	122.7	127.3	154.3	145.0	133.2	208.9
moreno_propro	100.0	103.2	107.6	104.9	105.8	109.5	99.0	102.0	121.4
moreno_train	100.0	100.6	110.4	116.3	121.0	212.9	164.1	178.1	109.5
munmun_digg_reply_LCC	100.0	99.7	108.4	103.2	102.6	102.7	102.8	101.7	101.4
opsahl-powergrid	100.0	106.0	103.4	126.3	128.9	161.0	142.6	118.0	364.2
power-eris1176	100.0	95.5	170.8	178.0	174.2	169.6	250.1	230.1	375.3
slashdot-zoo	100.0	100.8	103.5	113.2	112.1	102.2	103.3	101.9	99.4
subelj_jung-j	100.0	106.1	117.1	109.8	95.6	134.3	145.4	258.1	142.5
web-EPA	100.0	98.6	112.7	130.5	129.5	104.6	110.3	104.7	97.2
web-webbase-2001	100.0	120.1	93.0	142.3	153.4	150.2	189.7	185.9	892.9
Average	100.0	102.5	109.1	122.2	123.2	134.7	137.7	142.5	213.1

4 Conclusions

In this paper we present CoreGDM, a new algorithm for the Network Dismantling problem, based on the Graph Dismantling Machine (GDM) [19] framework and on CoreHD [40]. Extensive experiments on fifteen real-world networks show that the proposed algorithm provides a significant improvement over the original GDM formulation, and also over the other state-of-the-art algorithms, while also being more computationally efficient.

Acknowledgements. The Italian Ministry for Research and Education (MIUR) through Research Program PRIN 2017 (2017CWMF93), project 'Advanced Network Control of Future Smart Grids - VECTORS'.

References

1. Crime network dataset – KONECT, Apr 2017. http://konect.cc/networks/moreno_crime
2. Albert, R., Jeong, H., Barabási, A.L.: Error and attack tolerance of complex networks. Nature **406**(6794), 378–382 (2000). https://doi.org/10.1038/35019019

3. Arciprete, A., Carchiolo, V., Chiavetta, D., Grassia, M., Malgeri, M., Mangioni, G.: Geometric deep learning graph pruning to speed-up the run-time of maximum clique enumerarion algorithms. In: Cherifi, H., Mantegna, R.N., Rocha, L.M., Cherifi, C., Miccichè, S. (eds.) Complex Networks and Their Applications XI, pp. 415–425. Springer International Publishing, Cham (2023). https://doi.org/10.1007/978-3-031-21127-0_34
4. Boldi, P., Codenotti, B., Santini, M., Vigna, S.: UbiCrawler: a scalable fully distributed web crawler. Softw. Pract. Exp. **34**(8), 711–726 (2004)
5. Boldi, P., Rosa, M., Santini, M., Vigna, S.: Layered label propagation: a multiresolution coordinate-free ordering for compressing social networks. In: WWW, pp. 587–596 (2011)
6. Braunstein, A., Dall'Asta, L., Semerjian, G., Zdeborová, L.: Network dismantling. Proc. Natl Acad. Sci. **113**(44), 12368–12373 (2016). https://doi.org/10.1073/pnas.1605083113
7. CAIDA: Ipv4 routed /24 as links dataset. http://www.caida.org/data/active/ipv4_routed_topology_aslinks_dataset.xml
8. Carchiolo, V., Grassia, M., Longheu, A., Malgeri, M., Mangioni, G.: Exploiting long distance connections to strengthen network robustness. In: Xiang, Y., Sun, J., Fortino, G., Guerrieri, A., Jung, J.J. (eds.) IDCS 2018. LNCS, vol. 11226, pp. 270–277. Springer, Cham (2018). https://doi.org/10.1007/978-3-030-02738-4_23
9. Carchiolo, V., Grassia, M., Longheu, A., Malgeri, M., Mangioni, G.: Network robustness improvement via long-range links. Comput. Soc. Netw. **6**(1), 1–16 (2019). https://doi.org/10.1186/s40649-019-0073-2
10. Carchiolo, V., Grassia, M., Longheu, A., Malgeri, M., Mangioni, G.: Efficient node pagerank improvement via link building using geometric deep learning. ACM Trans. Knowl. Discov. Data (2022). https://doi.org/10.1145/3551642
11. Choudhury, M.D., Sundaram, H., John, A., Seligmann, D.D.: Social synchrony: predicting mimicry of user actions in online social media. In: Proceedings of International Conference on Computer Science and Engineering, pp. 151–158 (2009)
12. Cohen, R., Erez, K., ben Avraham, D., Havlin, S.: Breakdown of the internet under intentional attack. Phys. Rev. Lett. **86**(16), 3682–3685 (2001). https://doi.org/10.1103/physrevlett.86.3682
13. Coulomb, S., Bauer, M., Bernard, D., Marsolier-Kergoat, M.C.: Gene essentiality and the topology of protein interaction networks. Proc. R. Soc. B Biol. Sci. **272**(1573), 1721–1725 (2005)
14. Csardi, G., Nepusz, T.: The Igraph software package for complex network research. Int. J. Complex Syst. **1695**, 1–9 (2006). http://igraph.sf.net
15. De Nooy, W., Mrvar, A., Batagelj, V.: Exploratory Social Network Analysis with Pajek, vol. 27. Cambridge University Press, Cambridge (2011)
16. Ewing, R.M., et al.: Large-scale mapping of human protein-protein interactions by mass spectrometry. Mol. Syst. Biol. **3**, 89 (2007)
17. Fan, C., Zeng, L., Sun, Y., Liu, Y.Y.: Finding key players in complex networks through deep reinforcement learning. Nat. Mach. Intell. **2**(6), 317–324 (2020). https://doi.org/10.1038/s42256-020-0177-2
18. Fey, M., Lenssen, J.E.: Fast graph representation learning with PyTorch Geometric. In: ICLR Workshop on Representation Learning on Graphs and Manifolds (2019)
19. Grassia, M., De Domenico, M., Mangioni, G.: Machine learning dismantling and early-warning signals of disintegration in complex systems. Nat. Commun. **12**(1), 5190 (2021). https://doi.org/10.1038/s41467-021-25485-8

20. Grassia, M., Mangioni, G.: wsGAT: weighted and signed graph attention networks for link prediction. In: Benito, R.M., Cherifi, C., Cherifi, H., Moro, E., Rocha, L.M., Sales-Pardo, M. (eds.) Complex Networks & Their Applications X, pp. 369–375. Springer International Publishing, Cham (2022). https://doi.org/10.1007/978-3-030-93409-5_31
21. Guo, G., Zhang, J., Yorke-Smith, N.: A novel Bayesian similarity measure for recommender systems. In: Proceedings of International Joint Conference on Artificial Intelligence, pp. 2619–2625 (2013)
22. Hagberg, A.A., Schult, D.A., Swart, P.J.: Exploring network structure, dynamics, and function using NetworkX. In: Proceedings of the 7th Python in Science Conference (SciPy2008), pp. 11–15. Pasadena, CA USA, August 2008
23. Hamilton, W.L.: Graph representation learning. Synth. Lect. Artif. Intell. Mach. Learn. **14**(3), 1–159 (2020)
24. Han, J.D.J., Dupuy, D., Bertin, N., Cusick, M.E., Vidal, M.: Effect of sampling on topology predictions of protein-protein interaction networks. Nat. Biotechnol. **23**(7), 839–844 (2005)
25. Hayes, B.: Connecting the dots. can the tools of graph theory and social-network studies unravel the next big plot? Am. Sci. **94**(5), 400–404 (2006)
26. Holme, P., Kim, B.J., Yoon, C.N., Han, S.K.: Attack vulnerability of complex networks. Phys. Rev. E **65**, 056109 (2002). https://doi.org/10.1103/PhysRevE.65.056109, https://link.aps.org/doi/10.1103/PhysRevE.65.056109
27. Kipf, T.N., Welling, M.: Semi-supervised classification with graph convolutional networks (2016). https://doi.org/10.48550/ARXIV.1609.02907, https://arxiv.org/abs/1609.02907
28. Kunegis, J., Lommatzsch, A., Bauckhage, C.: The slashdot zoo: mining a social network with negative edges. In: Proceedings of International World Wide Web Conference, pp. 741–750 (2009). https://cc/kunegis/paper/kunegis-slashdot-zoo.pdf
29. Liao, R., et al.: Efficient graph generation with graph recurrent attention networks. In: Wallach, H., Larochelle, H., Beygelzimer, A., d' Alché-Buc, F., Fox, E., Garnett, R. (eds.) Advances in Neural Information Processing Systems, vol. 32. Curran Associates, Inc. (2019). https://proceedings.neurips.cc/paper/2019/file/d0921d442ee91b896ad95059d13df618-Paper.pdf
30. Matke, C., Medjroubi, W., Kleinhans, D.: SciGRID - An Open Source Reference Model for the European Transmission Network, vol. 2, July 2016. http://www.scigrid.de
31. Morone, F., Min, B., Bo, L., Mari, R., Makse, H.A.: Collective influence algorithm to find influencers via optimal percolation in massively large social media. Sci. Rep. **6**, 30062 (2016)
32. Peixoto, T.P.: The graph-tool python library. figshare (2014). https://doi.org/10.6084/m9.figshare.1164194, http://figshare.com/articles/graph_tool/1164194
33. Ren, X.L., Gleinig, N., Helbing, D., Antulov-Fantulin, N.: Generalized network dismantling. Proc. Natl. Acad. Sci. **116**(14), 6554–6559 (2019). https://doi.org/10.1073/pnas.1806108116, https://www.pnas.org/content/116/14/6554
34. Rossi, E., Chamberlain, B., Frasca, F., Eynard, D., Monti, F., Bronstein, M.M.: Temporal graph networks for deep learning on dynamic graphs. CoRR abs/2006.10637 (2020). arxiv.org:2006.10637
35. Rossi, R.A., Ahmed, N.K.: The network data repository with interactive graph analytics and visualization. In: AAAI (2015). http://networkrepository.com

36. Stumpf, M.P., Wiuf, C., May, R.M.: Subnets of scale-free networks are not scale-free: sampling properties of networks. Proc. Natl. Acad. Sci. U.S.A. **102**(12), 4221–4224 (2005)
37. Veličković, P., Cucurull, G., Casanova, A., Romero, A., Liò, P., Bengio, Y.: Graph attention networks. In: International Conference on Learning Representations (2018). https://openreview.net/forum?id=rJXMpikCZ
38. Šubelj, L., Bajec, M.: Software systems through complex networks science: review, analysis and applications. In: Proceedings of International Workshop on Software Mining, pp. 9–16 (2012)
39. Watts, D.J., Strogatz, S.H.: Collective dynamics of 'small-world' networks. Nature **393**(1), 440–442 (1998)
40. Zdeborová, L., Zhang, P., Zhou, H.J.: Fast and simple decycling and dismantling of networks. Sci. Rep. **6**(1), 37954 (2016). https://doi.org/10.1038/srep37954
41. Zhang, B., Liu, R., Massey, D., Zhang, L.: Collecting the Internet AS-level topology. SIGCOMM Comput. Commun. Rev. **35**(1), 53–61 (2005)

The Impact of a Crisis Event on Predicting Social Media Virality

Esra C. S. de Groot(✉), Reshmi G. Pillai, and Fernando P. Santos(✉)

Informatics Institute, University of Amsterdam,
Science Park 900, 1098XH Amsterdam, The Netherlands
ecsdegroot.uva@gmail.com, f.p.santos@uva.nl

Abstract. Understanding why specific information pieces become viral in online social networks is a fundamental step in governing social media platforms. While previous studies explored which user and network properties enable information virality, little is known about how offline crisis events impact these features. Here we investigate to what extent a crisis event impacts virality prediction models. We analyze Twitter data before and after the 2022 Russian invasion of Ukraine as a case study. We train and test statistical learning models on data before and after the invasion, evaluating which features related to content (e.g., existence of an URL), the user (e.g., number of previous tweets), or the network (e.g., degree centrality) are most relevant to predict virality. Tweets with more than 100 retweets are considered viral. We observe that the crisis event affects virality prediction: predictive accuracy is reduced when a model trained on data before the invasion is evaluated on data after the invasion—instead of being evaluated on data before the invasion. We observe that the crisis event leads to an increased fraction of viral tweets by users with fewer followers after the invasion. Moreover, we observe that, after the invasion, tweets containing URLs are more likely to become viral, implying an increased interest in website links, photos, and videos. Overall, our study reveals a shift in the importance of features underlying virality as a result of an offline crisis event.

Keywords: Social networks · Social media · Information diffusion

1 Introduction

Social media platforms are nowadays places where users can express themselves at an unprecedented scale [21]. The opportunity for sharing content on social media allows users to be exposed to a great amount of information, making social media platforms like Twitter, Instagram, and Facebook important assets when it comes to accessing real-time data [11]. More specifically, during crisis events like natural disasters [19], pandemics [17], or armed conflicts [29], information is often shared by people and authorities on social media platforms [22, 25, 26]. Not all of these messages become viral and are widely shared, though. Which user, content, or network features enable information to become viral? And how is the

A. S. Teixeira et al. (Eds.): CompleNet 2023, SPCOM, pp. 95–107, 2023.
https://doi.org/10.1007/978-3-031-28276-8_9

importance of such features impacted by crisis events? These are questions we approach in this paper.

Online viral information during crisis events has serious societal effects. Social media platforms provide a fast and effective manner to disseminate important information, which improves awareness and can offer safety information on how to react during and after a crisis event. On the downside, misinformation is often spread (un)intentionally by individuals or organizations during crisis events [17]. The spread of misinformation can lead to the lack of trust in reliable news sources and authorities, even facilitating conspiracy beliefs [3,9]. Machine learning models that predict when information spreads widely in an online social network can aid in tackling the spread of misinformation during crisis events and can be used to understand the drivers of information sharing by humans.

While prior studies investigated which features contribute to information going viral within an online social network [13,20], it remains unclear to what extent the influence of these features is impacted by real-life crisis events. As the online and offline worlds are blended in today's society, the relation between the underlying features and the possibility of a viral spread of information on social media is likely to be impacted by real-world events. In a broader context, machine learning prediction models can be affected by concept drift [6]: the relationship between input features underlying machine learning models and an output of interest can change over time. Therefore, some features underlying virality might become more or less important for model prediction during and/or after a crisis event, or the type of the relationship between a feature and virality might change. For example, the indegree (followers) of a user is often considered to be the most predictive feature for virality on Twitter [8,13,30], however, it is uncertain whether its relationship with virality might change due to a real-world event. As a result, this may impact virality prediction models' accuracy.

To study the effect of a crisis event on the virality prediction model in online social networks, we focus on Twitter as a social media platform, together with the invasion of Russia in Ukraine (24-02-2022) as a crisis event. Twitter is chosen because it is a widely used platform, often used to study information spread [7,8,13,15,17,19,20,29,30]. We retrieved tweets posted before the invasion (February 21 to 24, 2022) and right after the invasion (February 24 to 25, 2022).

We believe this is a unique dataset as it allows to study the influence of real-world events on feature dynamics underlying virality on Twitter, while still analyzing tweets about the same broader topic. Contrarily, in prior literature, only tweets posted during or right after the crisis event [19], or users that tweeted about the event [7], are investigated. The latter is achieved by investigating tweets about another topic prior to the event. Furthermore, the dataset provides a timeserving perspective of how a real-life crisis event impacts the influences of content, network, and user features underlying the virality of tweets.

We perform a descriptive analysis to investigate to what extent content, network, and user features underlying a tweet virality prediction model are impacted by a real-world crisis event. The characteristics and performance of three binary virality classification models are analyzed. One model is trained on data before the invasion (before model), one model on data after the invasion (after model), and one context-ignorant control model on before and after data (ignorant model). Using these models, we explore (i) whether the model performances are context-dependent, (ii) how important and what type of relationship content, network, and user features have within the model, and (iii) whether the relationships between the features and virality differ before and after the event.

2 Related Work

2.1 Defining Virality

Providing a proper definition for virality is difficult as this concept can be context-specific, depending on specific datasets or topics. One simple way to define virality is to consider the number of retweets and define a threshold for the minimum number of retweets to classify a tweet as viral [13]. Another common measure is structural virality [10,31], which also considers the pattern of retweet cascades. Given the difficulty of building a retweet cascade from the functionality provided by Twitter's current API, we use the number of retweets as a measure of virality. We consider that a tweet is viral if it is retweeted a number of times above a certain threshold. This is a similar approach to the method of Jenders et al. and Neppalli et al. [13,19] (see Sect. 3.1 for details).

2.2 Crisis Events on Twitter

During crisis events, social media is often used as a source of information about the event [22,25,26]. The existing literature is often focused on the dynamics of features during or after the crisis event, without taking into account the state before the event [17,19,29].

The study by Neppalli et al. considers a model that predicts how likely a tweet is to be retweeted more than a specific threshold (with 20 as the best-performing threshold), using 7.5M initial tweets about hurricane Sandy [19]. Both user and network features were found to be important predictors of virality. However, there was no distinction made between which user features were the most meaningful. Another study found the number of followers (indegree) to be the most important predictor when predicting retweet chain size using data from the Japanese Earthquake and Tsunami [24]. Unsurprisingly, also within non-crisis events, the number of followers (indegree) is often found to be the strongest predictor for viral information diffusion [8,13,30]. Therefore, we expect the number of followers and the number of listed users (a related network feature) to be among the most important predictors.

While the study by Buntain et al. did include a comparison between before and after crisis event data [7], the tweets did not include a similar context. The study analyzed the effect of three crisis-event Twitter datasets on activity and social interactions on Twitter: the 2013 Boston Marathon Bombing, the 2014 Sydney Hostage Crisis, and the 2015 Charlie Hebdo Shooting.

Tweets were scraped two weeks before and after the events, using the keywords *Boston, Sydney, Paris,* and *Hebdo.* The work of Buntain et al. differs from ours in several dimensions: Buntain et al. do not focus on any predictive task; moreover, the study focuses on understanding how different topics, containing similar keywords, are discussed by the same users (one crisis and one non-crisis), rather than studying the influence of an event on an already discussed topic. For instance, tweets containing the word 'Boston' could describe the weather when posted before the event, but after the event, the same keyword mainly addresses the bombing. In the current study, the tweets from the before invasion dataset already address the situation between Russia and Ukraine.

The results of Buntain et al. are interesting with respect to our expectations of what features underlying virality are important within a crisis context. Their study notes that the number of relevant tweets, retweets, hashtags, and URLs increases significantly after the events; the topic itself goes viral on Twitter. Moreover, accounts from the police, government, and news organizations become an important means to obtain prompt information. The latter result suggests that the influence of some user features might be impacted by crisis events.

Another interesting study about the war in Gaza between Palestine and Israel highlights that, next to the usual users of interest (e.g. politicians, journalists, and activists), a new group of users get an important role in the dissemination of information on Twitter: witnesses [29]. This group of users is becoming an interesting and important source of information during crises, such as wars, because they tweet about their first-hand experiences. Importantly, this class of users is not necessarily central in the network according to degree centrality, i.e., the number of followers or the number of listed users (how many users have added you to their list). Therefore, we expect in our current study that the network features are impacted by the crisis event.

3 Methodology

3.1 Datasets

Crisis Event Data Extraction. Twitter data was extracted using the historical Twitter API and Tweepy [27], between the 14th and the 16th of March 2022. All scraped tweets were written in English because it is the most used language on Twitter [12], and the most convenient for sentiment analysis. To avoid duplicate tweets, all collected tweets were originally written tweets and did not contain any replies, retweets, or quoted tweets. Keywords used to scrape tweets related to the invasion of Russia in Ukraine were: *Ukraine*, *#Ukraine*, *Russia*, *#Russia*, *Putin*, *#Putin*, *Zelensky*, *#Zelensky*, and *#UkraineRussianWar*. This means that all collected tweets, posted by publicly available user accounts on Twitter, contained one or more of these keywords. These keywords are chosen because of their likelihood to represent content related to the tensions/invasion between Russia and Ukraine and/or because they were trending.

On February 24 2022, Putin announced the invasion of Ukraine on national television a little before 03:00 UTC, after which explosions were reported in

Kyiv, Kharkiv, Odesa, and the Donbas region [1,28]. Therefore, tweets posted before the invasion were scraped using a historical time frame from 02:30 UTC on 21-02-2022 up to 02:30 UTC on 24-02-2022: the before data set. Tweets posted after the invasion were scraped using a historical time frame from 03:00 UTC on 24-02-2022 up to 02:30 UTC on 25-02-2022: the after data set. As 73% of all retweets already occur within 2 h after posting the original tweet [32], the time gap of 18-23 days between posting and scraping the tweets should be enough to capture an accurate retweet number for each tweet. The Ukraine before data consisted of the tweet and user features of 697,498 initial tweets, written by 276,612 unique users. For the Ukraine after data 1,365,732 initial tweets were scraped, written by 783,812 unique users.

Table 1. Description of all content/tweet, network, and user features, and the target feature of the before invasion data (21-02-22 to 24-02-22) and after invasion data (24-02-22 to 25-02-22). The author refers to the user that originally posted the tweet.

Type	Feature	Description
Content/ Tweet	emoji_count	Number of emojis used in the tweet
	sentiment	Compound sentiment score between 1 (positive) and -1 (negative) (NLTK's Vader Sentiment analysis [5])
	urls_count	Number of URLs in the tweet
	mention_count	Number of mentions in the tweet
	hashtag_count	Number of hashtags in the tweet
	tweet_char_length	Number of characters in the tweet
Network	log_followers	Logarithm of the number of users that follow the author
	log_following	Logarithm of the number of users that the author follows
	log_listed	Logarithm of the number of lists the author is part of (listed by other users)
User	log_tweetcount	Logarithm of the number of historical tweets of the author
	account_age_y	Number of years that the account of the author exists
	verified	1 if the user is verified
	sex_generalized	1 if a male, -1 if female, and 0 if unknown (gender-guesser [14])
	organization	1 if most likely an organization (if $verified = 1$ and $sex\_generalized = 0$)
Target	viral	1 if the number of retweets is more than 100

Data Description. All content, network, and user features can be found in Table 1 together with their descriptions. Logarithmic transformations were applied to some network and user features to make them suitable for linear models. The number of followers and the number of listed users are both metrics of the users' indegree, and the number of following of the outdegree. On Twitter, lists are used to prioritize and organize your timeline. The listed feature describes the number of users that put the author of the tweet in their list.

The binary target variable $viral = \{0, 1\}$ is created using the number of retweets of the tweet provided by the Twitter API. A tweet is marked as viral if the number of retweets is higher than 100. This threshold of 100 retweets is based on two different considerations: on one hand, the threshold should be high enough to capture true viral tweets. More retweets of a tweet mainly mean a higher virality and usually result in a better distinction from non-viral tweets [13]. On the other hand, a higher threshold for virality results in a smaller sample of viral tweets, and therefore a larger class imbalance. These two aspects led to the decision to have a threshold of 100 retweets. We assume that this number is still high enough to be considered viral while not impacting the model performance too drastically by creating a large class imbalance.

Splitting the Data. All datasets are randomly split into a train and test set (respectively 85% and 15%). The goal of the study is to analyze the effect of the invasion. Therefore, the three train and test datasets need to be equally sized with an equal class distribution to make the models and their performance comparable. The train and test datasets from the after data and combined data (for the ignorant model) are randomly down-sampled using *Imbalanced-learn* [16] to fit the dataset size and class balance of the smallest dataset: the before dataset. This resulted in a final train set of 592,873 tweets (6376 viral tweets) and a final test set of 104,625 tweets (1109 viral tweets) for all datasets.

3.2 Experimental Setup

To study the impact of the invasion on the dynamics of the features underlying tweet virality, three final binary classification models are trained to predict tweet virality. The before model is trained on the Ukraine before data. The after model is trained on the Ukraine after data. The ignorant model serves as a control model that is trained on both before and after data. Therefore, the ignorant model is not aware of the crisis event. If the invasion impacts the virality prediction model, it is expected that the performance of the before model is better when tested on the before test set, compared to the after test set. Likewise, it is expected that the performance of the after model is better when tested on the after test set, compared to the before test set. Moreover, it is expected that the ignorant model performs worse than the two specified models because it ignores the period when data was created.

Model Training. To come up with the final models, four different supervised machine learning classification models are trained: a logistic regression model (LG), a random forest classifier (RF), an extreme gradient boosting classifier (XGB), and a multi-layer perceptron classifier (MLP). The first fit of these models, without hyperparameter tuning or taking class imbalance into account, served as the baseline models. The LG model is chosen because it is relatively simple to interpret the relationships between the features and the target variable. However, the LG model is not able to capture non-linear relationships, unlike the MLP, RF, and XGB classifiers. The MLP classification is based on a neural network and the RF and XGB are both based on an ensemble of decision trees, which makes them more robust to outliers and less prone to overfitting.

To reduce the risk of overfitting, the models were trained using 5-fold cross-validation using Scikit-learn version 1.0.2 [23]. Class balances were taken into account when making the splits. Throughout the entire process of model training, the macro averaged F1 score is used to measure model performance. For the LG and MLP classifiers, predictive features of the data were first scaled using a standard scaler fitted on the train data (to prevent data leakage to the validation set). The RF model turned out to be the best model for the before and after model. Therefore, the results (Sect. 4) describe only the RF classifier models.

As the class distribution is highly imbalanced, multiple resampling techniques were tried to improve model performance. The sampling strategies include random oversampling, random undersampling, SMOTE, and a bound method (setting a maximum number of retweets for the non-viral class). The bound method was found to be the best resampling method for our data. Afterward, hyperparameters were tuned for the models using a cross-validated grid search (see Table 2 for the final settings). Moreover, the same type of final model and resampling strategy is chosen for the ignorant model to save some computational time, as the type of train data is the same as the before and after train data. Afterward, the best ratio/bound method was searched for the ignorant model and its hyperparameters were tuned as well (see Table 2).

Model Evaluation. The performance of the final models is evaluated using the averaged macro F1 score because the accuracy of both classes is equally important. To make a statistical comparison of the hypotheses, the test sets are bootstrapped. For each original test set, 1000 bootstrapped samples of the size of the original test set are created by taking random instances from the original test set with replacement. For each bootstrapped sample, the macro F1-score is calculated, resulting in a macro F1 probability distribution. The before and after model are both tested on the before and after test set. The ignorant model is only tested on the combination test set.

The bootstrapped confidence intervals of the differences in the macro F1 distributions are used to decide whether the observed differences are significant. The significance level used is $\alpha = 0.05$. However, because the F1 distributions are tested twice (e.g. for the before model: 1) if accuracy is different for before and after test data, and 2) the difference between the before model and ignorant model with corresponding test data), a correction for multiple testing is done using a Bonferroni correction, resulting in a new significance level of $\alpha' = \frac{0.05}{2} = 0.025$ (a confidence interval of 97.5 %). If the confidence interval of the difference in the macro F1 distribution overlaps zero, the difference is not significant.

Feature Importance and Relationships. To investigate the changes in feature importance, we looked at the permutation importance for each feature for

Table 2. Final hyperparameters, resampling bound, and F1 train scores for the random forest models. All models outperformed a majority baseline model (always predicting non-viral, $F1 = 0.50$).

	n estimators	Min samples_split	Max depth	Max features	Criterion	bound (ratio)	F1 train (SD)
Before model	150	5	40	sqrt	Entropy	25 (0.0111)	0.752 (0.004)
After model	150	2	20	sqrt	gini	10 (0.0114)	0.736 (0.005)
Ignorant model	150	5	40	sqrt	entropy	25 (0.0111)	0.736 (0.004)

both before and after models [2,18], using the implementation in Scikit-learn 1.0.2 [23]. The permutation importance describes the average drop in macro F1 score as a result of permuting the feature with random values. Because the aim of using the permutation importance in this study is to get a better understanding of how the models work and on what features the predictions are based, the permutation importance is computed on the train set [18]. The permuted importances ($n = 30$ per feature) are divided by the total macro F1 score of the trained fit of the model and multiplied by 100 to get the percentage of macro F1 drop.

Furthermore, visualizations are made to explain the relationship of a feature with virality and its change after the event. Based on prior research, we focused in this paper on the number of followers [7,29] and the number of URLs [7].

4 Results

4.1 Model Evaluation

The F1 scores of the final random forest classifier model trained with before/after event data and tested with before/after test data can be found in Table 3, together with the F1 score of the ignorant model. For the ignorant model, we trained and tested the model on the whole dataset, ignoring the period during which the tweets were produced. For the before model, the 97.5% confidence interval of the difference in F1 score when tested on the before and after test set does not overlap zero ($BB - BA = [0.0111, 0.0590]$). For the after model, the 97.5% confidence interval of the difference in F1 score when tested on the before and after test set also does not overlap zero ($AA - AB = [0.0144, 0.0530]$). Therefore, both models perform significantly better on their

Table 3. Macro F1 scores (SD) on the bootstrapped test sets. It can be seen that the RF models perform better when trained and tested in data produced in the same period.

	Before test	After test	Combined test
Before model	0.7503 (0.0066)	0.7139 (0.0076)	-
After model	0.6962 (0.0055)	0.7290 (0.0065)	-
Ignorant model	-	-	0.7378 (0.0071)

corresponding test set: the crisis event affects the possibility that information virality in an online social network is accurately predicted. In contrast, the ignorant model does not perform significantly worse than both specified models ($BB - CC = [-0.0099, 0.0338]; AA - CC = [-0.0288, 0.0135]$).

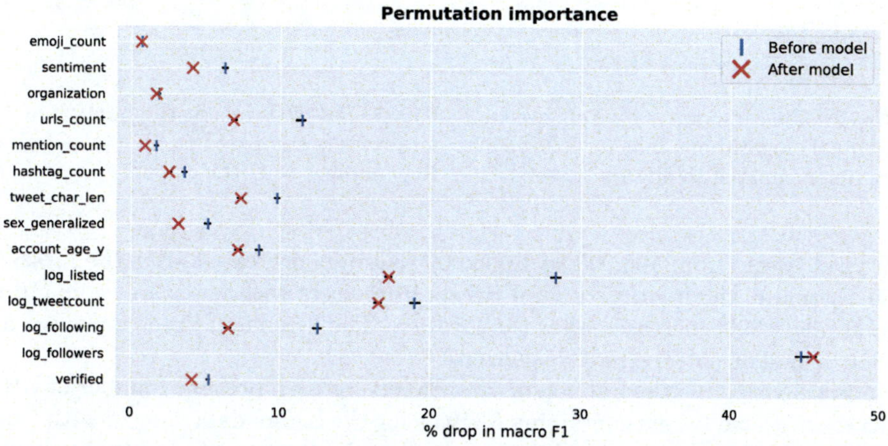

Fig. 1. The mean permutation importance for each feature of the before model (blue stripe) and after model (red cross). The permutation importance is measured by the percentage of macro F1 drop after randomly permuting the feature, compared to the total macro F1 score of the model. A bigger drop means higher importance. The spread of the minimum and maximum values for each importance are shown with a black line. If the line is (almost) unnoticeable, the extreme values are close to the mean. (Color figure online)

4.2 Feature Importance and Relationships

The distributions of the relative permutation importance are visualized in Fig. 1. The feature with the highest mean permutation importance (i) for the before (b) and after (a) model is the number of followers ($i_b = 44.7837 \pm 0.0337, i_a = 45.6161 \pm 0.0205$). The second most important feature is the number of listed users ($i_b = 28.3316 \pm 0.0394, i_a = 17.2663 \pm 0.0214$), this feature also shows the biggest difference in importance. The third most important feature is the number of tweets from the user ($i_b = 18.9824 \pm 0.0311, i_a = 16.5600 \pm 0.0206$), followed by the number of following ($i_b = 12.5535 \pm 0.0277, i_a = 6.5795 \pm 0.0172$) and the URLs count ($i_b = 11.4582 \pm 0.0327, i_a = 6.8701 \pm 0.0197$). Except for the number of followers, we see a slight reduction in permutation importance in the after model. Taken together, this shows that the indegree network features (number of followers and listed users) are overall important predictors for virality.

While the importance of the number of followers is relatively similar for both models, its relationship with virality looks different before and after the invasion. The fraction of viral tweets per number of followers is shown in Fig. 2, the peak

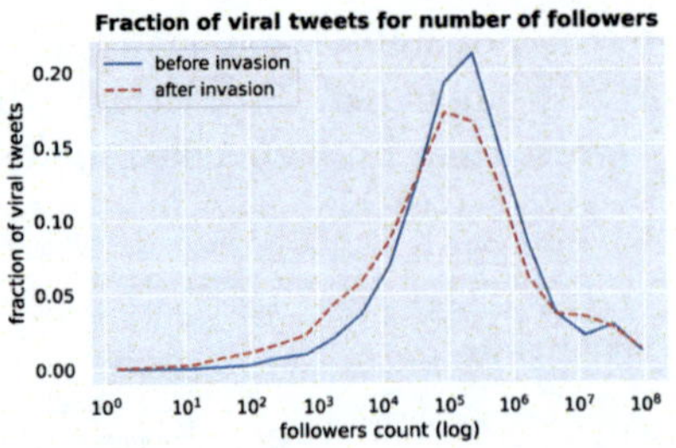

Fig. 2. The fraction of viral tweets within 20 logarithmic bins for followers counts, compared to all viral tweets.

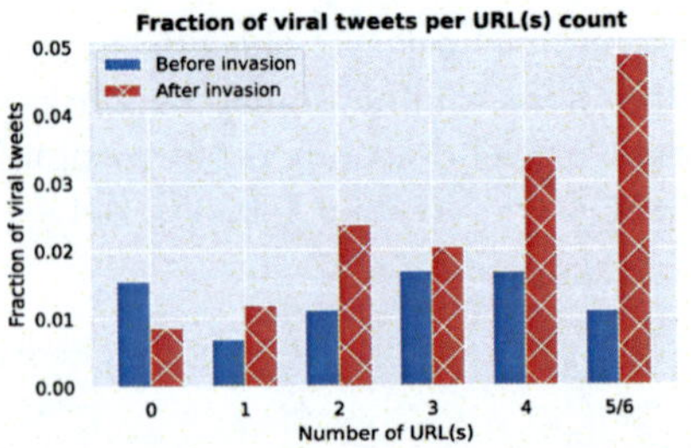

Fig. 3. Per URLs count the fraction of viral tweets compared to all tweets for that number of URLs count.

in viral tweets from 100,000 to 1,000,000 followers decreased after the invasion. Furthermore, the fraction of viral tweets from users with fewer followers (10 up to around 1000) increased after the invasion. The listed and tweetcount features show a similar pattern (not visualized).

Fig. 3 visualizes the fraction of viral tweets for each possible number of URLs (as there are no tweets containing 6 URLs for the before data, tweets with 5 or 6 URLs are taken together). After the invasion, the fraction of viral tweets having at least one URL increased compared to before the invasion.

5 Discussion and Conclusion

The aim of this study was to investigate to what extent content, network, and user features underlying a tweet virality prediction model are impacted by a real-world crisis event: the Russian invasion of Ukraine. The performance of the random forest classifier model trained on data before the invasion (before model) significantly decreased when tested with data after the invasion, compared to test data from before the invasion. Likewise, the after model performed significantly better on the after test data, compared to the before test data. The change in model predictive capabilities was expressed by an overall reduction of the feature permutation importance, except for the number of followers. Taken together, we can conclude that the features underlying a virality prediction model can be affected by a crisis event, impacting model accuracy.

As expected based on prior studies [7,24,29], the most important and most impacted features were mainly the network centrality features. The number of followers and the number of listed users were both key features in virality prediction. While we see that the number of followers preserves its predictive power (feature importance) before and after the invasion (Fig. 1), its relationship with virality is changed: we see an increased fraction of viral tweets after the invasion for users with fewer followers (Fig. 2). This effect can possibly be explained by a shift in the importance and relevance of specific users: information posted by witnesses (not necessarily with a high indegree) who first-hand experienced the crisis event is an important source of information [29]. Another user feature, the number of previous tweets, was also an important predictor for virality, representing prior user engagement on Twitter.

While the feature importance for the number of URLs decreases after the invasion, there are some interesting changes seen in the visualization of the number of URLs before and after the invasion (see Fig. 3). After the invasion, the fraction of viral tweets increases when the number of URLs increases. As an example, from all tweets containing 4 URLs, a higher fraction is viral after the invasion. The increased fraction of viral tweets for tweets containing more URLs after the invasion may represent a stronger interest in information coming from websites, photos, and videos after a crisis event, in line with prior research [7].

Unexpectedly, the specified models did not outperform the ignorant combined model, probably as the ignorant model is trained to generalize well over two different contexts (before and after the invasion). Therefore, future research could address how to decide on the best trade-off for data inclusion to adapt a prediction model to quick and abrupt changes caused by a crisis event (e.g. by assigning weights to more recent data).

There are some limitations that could have impacted the generalizability of our results. Firstly, as all tweets were in English, our results should be interpreted from an international level of communication. Possibly, results could yield different conclusions when only Ukrainian/Russian tweets are used. Furthermore, more research is required to investigate whether the current conclusions hold for other crisis events (e.g. a natural disaster or pandemics). Our results show that a social media virality prediction model can be impacted by a specific crisis event, suggesting future research to evaluate how these observations generalize to events of a different scale and nature.

As the indegree network features (number of followers and listed users) were key predictors for virality, we suggest that future works investigate the impact of a crisis event on the network structure of virality itself. This can eventually consider higher-order networks: the Twitter lists of users can be considered hyperedges, i.e., a generalization of (pairwise) edges as the connection between larger sets of nodes [4]; in this regard, features such as order or hyperdegree might be relevant features to infer virality. Future research could also analyze whether viral tweets frequently remain within clusters of similar users and whether the observed level of inter-community outreach is impacted after a crisis event.

Data and Code Availability. The data and code used for this study are available on Github[1]. Due to Twitter's limitations, only tweet ids are shared.

Acknowledgements. We acknowledge the University of Amsterdam - Master Programme Information Studies for creating the conditions to perform this research and for financially supporting this publication. Furthermore, we acknowledge the anonymous reviewers for the excellent recommendations to improve this article.

[1] https://github.com/sudegroot/CompleNet_virality_social_network.

References

1. Full text: Putin's declaration of war on Ukraine. The Spectator, February 2022. https://www.spectator.co.uk/article/full-text-putin-s-declaration-of-war-on-ukraine/. Accessed 13 Mar 2022
2. Altmann, A., Toloşi, L., Sander, O., Lengauer, T.: Permutation importance: A corrected feature importance measure. Bioinformatics. **26**(10), 1340–1347 (2010). https://doi.org/10.1093/bioinformatics/btq134
3. Ball, P., Maxmen, A.: The epic battle against coronavirus misinformation and conspiracy theories. Nature. **581**(7809), 371–374 (2020)
4. Battiston, F., et al.: Networks beyond pairwise interactions: structure and dynamics. Phys. Rep. **874**, 1–92 (2020)
5. Bird, S., Klein, E., Loper, E.: Natural Language Processing with Python: Analyzing Text with The Natural Language Toolkit. O'Reilly Media, Inc., Sebastopol (2009)
6. Brzezinski, D., Stefanowski, J.: Reacting to different types of concept drift: the accuracy updated ensemble algorithm. IEEE Trans. Neural Netw. Learn. Syst. **25**(1), 81–94 (2014). https://doi.org/10.1109/TNNLS.2013.2251352
7. Buntain, C., Golbeck, J., Liu, B., Lafree, G.: Evaluating public response to the Boston marathon bombing and other acts of terrorism through Twitter. In: Proceedings of the Tenth International AAAI Conference on Web and Social Media (2016)
8. Bunyamin, H., Tunys, T.: A comparison of retweet prediction approaches: the superiority of random forest learning method. Telecommun. Comput. Electron. Control. **14**(3), 1052 (2016). https://doi.org/10.12928/telkomnika.v14i3.3150
9. Enders, A.M., et al.: The relationship between social media use and beliefs in conspiracy theories and misinformation. Polit. Behav. 1–24 (2021). https://doi.org/10.1007/s11109-021-09734-6
10. Goel, S., Anderson, A., Hofman, J., Watts, D.J.: The structural virality of online diffusion. Manage. Sci. **62**(1), 180–196 (2016). https://doi.org/10.1287/mnsc.2015.2158
11. Haewoon, K., Changhyun, L., Hosung, P., Sue, M.: What is Twitter, a Social Network or a News Media? In: Proceedings of the 19th International Conference on World Wide Web, p. 1365 (2010)
12. Hong, L., Convertino, G., Chi, E.: Language matters in Twitter: a large scale study. In: Proceedings of the International AAAI Conference on Web and Social Media, vol. 5, pp. 518–521 (2011)
13. Jenders, M., Kasneci, G., Naumann, F.: Analyzing and predicting viral tweets. In: WWW 2013 Companion - Proceedings of the 22nd International Conference on World Wide Web, pp. 657–664. Association for Computing Machinery (2013). https://doi.org/10.1145/2487788.2488017
14. Michael, J.: 40000 names (2017)
15. Kupavskii, A., et al.: Prediction of retweet cascade size over time. In: Proceedings of the 21st ACM International Conference on Information and Knowledge Management, pp. 2335–2338 (10 2012)
16. Lemaître, G., Nogueira, F., Aridas char, C.K.: Imbalanced-learn: a python toolbox to tackle the curse of imbalanced datasets in machine learning. J. Mach. Learn. Res. **18**, 1–5 (2017). http://jmlr.org/papers/v18/16-365.html.
17. Mirbabaie, M., Bunker, D., Stieglitz, S., Marx, J., Ehnis, C.: Social media in times of crisis: learning from Hurricane Harvey for the coronavirus disease 2019 pandemic response. J. Inf. Technol. **35**(3), 195–213 (2020). https://doi.org/10.1177/0268396220929258

18. Molnar, C.: 8.5 Permutation Feature Importance. In: Interpretable Machine Learning. A guide for Making Black Box Models Explainable, pp. 193–203 (2019)
19. Neppalli, V.K., et al.: Retweetability Analysis and Prediction during Hurricane Sandy. ISCRAM (2016)
20. Nesi, P., Pantaleo, G., Paoli, I., Zaza, I.: Assessing the reTweet proneness of tweets: predictive models for retweeting. Multimed. Tools Appl. **77**(20), 26371–26396 (2018). https://doi.org/10.1007/s11042-018-5865-0
21. Neubaum, G., Krämer, N.C.: Opinion climates in social media: blending mass and interpersonal communication. Human Commun. Res. **43**(4), 464–476 (2017). https://doi.org/10.1111/hcre.12118
22. Oh, O., Agrawal, M., Rao, H.R.: Community intelligence and social media services: a rumor theoretic analysis of tweets during social crises. Technical Report. 2 (2013)
23. Pedregosa, F., et al.: Scikit-learn: machine learning in python. J. Mach. Learn. Res. **12**, 2825–2830 (2011). http://scikit-learn.sourceforge.net
24. Remy, C., Pervin, N., Toriumi, F., Takeda, H.: Information diffusion on twitter: everyone has its chance, but all chances are not equal. In: Proceedings - 2013 International Conference on Signal-Image Technology and Internet-Based Systems, SITIS 2013, pp. 483–490 (2013). https://doi.org/10.1109/SITIS.2013.84
25. Reuter, C., Hughes, A.L., Kaufhold, M.A.: Social media in crisis management: an evaluation and analysis of crisis informatics research. Int. J. Human-Comput. Interact. **34**(4), 280–294 (2018). https://doi.org/10.1080/10447318.2018.1427832
26. Reuter, C., Kaufhold, M.A., Schmid, S., Spielhofer, T., Hahne, A.S.: The impact of risk cultures: citizens' perception of social media use in emergencies across Europe. Technol. Forecast. Soc. Change. **148**, 1–7 (2019). https://doi.org/10.1016/j.techfore.2019.119724
27. Roesslein, J.: Tweepy: Twitter for Python! (2020). https://Github.com/Tweepy/Tweepy
28. Sheftalovich, Z.: Battles flare across Ukraine after Putin declares war, February 2022
29. Siapera, E., Hunt, G., Lynn, T.: #GazaUnderAttack: Twitter, Palestine and diffused war. Inf. Commun. Soc. **18**(11), 1297–1319 (2015). https://doi.org/10.1080/1369118X.2015.1070188
30. Suh, B., Hong, L., Pirolli, P., Chi, E.H.: Want to be retweeted? Large scale analytics on factors impacting retweet in Twitter network. In: Proceedings - SocialCom 2010: 2nd IEEE International Conference on Social Computing, pp. 177–184 (2010). https://doi.org/10.1109/SocialCom.2010.33
31. Vosoughi, S., Roy, D., Aral, S.: The spread of true and false news online. Science. **359**(6380), 1146–1151 (2018). https://doi.org/10.1126/science.aap9559
32. Yin, H., Yang, S., Song, X., Liu, W., Li, J.: Deep fusion of multimodal features for social media retweet time prediction. World Wide Web **24**(4), 1027–1044 (2020). https://doi.org/10.1007/s11280-020-00850-7

Evaluating the Bayesian MRP Network Model for Estimating Heterogeneity in (Age-Stratified) Contact Patterns from Highly Selective Samples

Ramona Ottow(✉)

Universitat Pompeu Fabra, Barcelona 08005, Spain
ramona.ottow@upf.edu

Abstract. When estimating social contact patterns from biased or censored data sets, methods that are able to address this issue while simultaneously capturing the intensity as well as contact structure in their estimates are needed. This simulation study aims to examine the performance of Bayesian Multilevel Regression with Poststratification and its two extensions that utilize local network information to estimate age-stratified mean number of in-home contacts. I compare the effect of including network information as well as of censored and biased sample data on the accuracy and examine in detail the ability to capture heterogeneity in (age-stratified) contacts.

Keywords: Bias · Heterogeneity · Multilevel regression and poststratification · Reciprocity · Social network

1 Introduction

Various approaches for estimating social contact patterns have been revisited or developed during the Covid-19 pandemic. Among these, Bayesian Multilevel Regression with Poststratification (MRP) is currently the most promising for population-level, age- and location-specific estimates [1,9]. In this work, I consider Bayesian MRP together with two extensions that leverage network information (reciprocity) of the underlying social networks.

The focus is on the examination of the methods' abilities, first, to accurately and precisely estimate contact intensity and contact patterns between different age groups, and second, to preserve the heterogeneous or homogeneous nature of the underlying social networks. I examine the effect of the inclusion of reciprocity in the MRP method as well as the effect of application in highly selective samples.

Knowledge about number of contacts between individuals and their age- and location-related contact patterns can contribute to preventing or mitigating the spread of infectious diseases like Covid-19. Mathematical modeling of disease dynamics can improve our understanding and predictions when it incorporates better information on these contact patterns [4,11].

A. S. Teixeira et al. (Eds.): CompleNet 2023, SPCOM, pp. 108–119, 2023.
https://doi.org/10.1007/978-3-031-28276-8_10

An evaluation of the methods, in particular the extensions, in the context of social contact patterns has not been done yet. The reciprocity constraints are applied in many papers on contact estimation methodologies [1,2,6] in combination with different approaches, but their effect on the final estimates has not yet been investigated.

The aim of this work is to evaluate the method's ability to predict age-stratified population-level mean number of in-home contacts for the Spanish population under a set of sampling scenarios.

I conduct a simulation study and utilize typical performance measures as well as network measures for the purpose of evaluating the methods.

The key finding is that the inclusion of reciprocity constraints can introduce bias to the results, but also improve the accuracy, in particular in the case of highly selective samples.

Furthermore, I find that the methods are able to account for biases in the samples in the estimation process, but still do not completely capture existing contact pattern between different age groups. In particular, the methods fail to capture the structure of the contacts completely, in the case of fully missing information on contacts within a specific age groups. In this study, I find this result for the age groups of children and adolescents for which the data is often missing in surveys due to ethical reasons. In this case, researcher should be careful in using this estimates in their work or rely on other methods, like the scaling method from [6] that uses a scaled version of contact information from other sources, while preserving the contact structure, as a imputation for the missing contacts.

2 Methodology

2.1 Objectives

The aim of this work is to evaluate the method's ability to predict age-stratified population-level mean number of in-home contacts for the Spanish population under a set of sampling scenarios.

2.2 Simulation Study

Simulation studies offer a way to compare statistical methods in terms of performance [7]. I start by generating a "true" population from an existing data set (Sect. 2.3). The next step is to generate samples by repeated resampling with replacement and applying the methods to each sample in order to examine the distribution of the estimates in comparison to the "true"population values.

The number of simulations is determined based on the precision of the estimation of my most important performance measure, the percentage bias (detailed description in Sect. 2.7). Following the recommendation of [7,10], I conduct a small test simulation with 25 repetitions and obtain a rough estimate of 0.5 for the estimate's standard deviation. Assuming the standard deviation is smaller or equal to 0.5 and demanding further a Monte Carlo SE of percentage bias of less than 0.05 yields a minimum requirement of 100 repetitions.

In order to avoid dependencies between the simulations and also to create reproducible work, I set a random seed once at the beginning of the algorithm and store as well as load the states of the random number generator, accordingly to the stimulation streams, before and after each repetition [7].

Additionally, I include a consistency check to identify outliers or pattern of the estimates between the simulations as well as (programming) errors. I examine the variation of the (age-stratified) estimates around their (age-stratified) mean value throughout the simulations. This check was realized by examination of different visualizations of the results.

The simulation study consists of four main steps, which I will present in detail in the next four sections.

2.3 Data Generating Mechanism

The "true" population in this study is the most realistic and representative data on contacts that is available for Spain at the moment of conducting the study. Using a realistic data set has the advantage that it naturally contains all structure and information and these are not artificially imposed by the researcher. The disadvantage is that the true data generation mechanism is unknown.

I investigate the methods' behaviour in different sampling scenarios, especially in settings where the amount of missing data is large. I use data from Spain's Labour Force Survey ("Economically Active Population Survey")[1], which is a large representative sample for Spain's population.

I use a subset of the socio-demographic data that contains, among other, gender, occupation, first- and second-level political and administrative division (autonomous communities and provinces), age group, and education.

The data set does not contain information on number of contacts, but it includes the number of household members. I assume in this work that a person has contact with each member of its household and therefore use the person's household members as a proxy for the in-home contacts of this person.

The ground truth data is constructed by calculating the number of in-home contacts for each of the 164,536 individuals in the data set. For each one a scaling factor is provided, which explains how many individuals it represents in the population. Consequently, I obtain representative in-home contact information for all 46.52 million non-institutionalized citizens in the country.

Descriptive statistics are provided in the Appendix A.

2.4 Sampling Mechanisms

I consider three sampling scenarios for this simulation study.

First, a censored simple random sample with completely missing data on contacts from, within and between the young age groups (under 20 years old).

---

[1] Labor Force Survey (Encuesta de población activa), https://www.ine.es/. I use data from the year 2021, first quarter, which coincides the best with the additionally used data from the third wave of the Distancia Covid Survey.

This is a "good, extreme" sample, because a lot of data is missing, but very good data within and between the older age groups is available.

The second sample is a censored simple random sample (CSRS) with missing data on contacts from the young age groups to all other age groups, but with information from older to the younger age groups. It is oriented on surveys, where often due to ethical reasons only adults participate and therefore provide information on their contacts with non-adults.

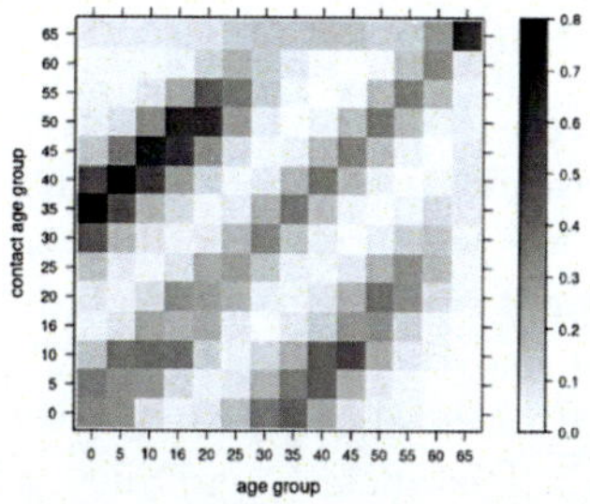

Fig. 1. The age-stratified contact matrix displaying the mean number of total contacts by age group for the ground truth data set. The diagonal line and its parallels suggest a clustering of contacts for specific age group pairs.

The third case is motivated by the large online Distancia-Covid Survey[2] which was conducted to better understand changing patterns of human mobility and social contacts in Spain during the Covid-19 pandemic [9]. This data is non-representative of the Spanish population due to self-selection bias.

I generate samples with similar statistics with the aim of learning about how the participation biases might affect the outcome and how well the methods work under extremely selective samples. Equally to the CSRS these Distancia-Covid samples (DCS) lacking information on contacts within children and adolescents and from them to adults.

The sample size is determined based on the outcome of the Distancia-Covid Survey. I sample 3131 observations in each repetition, which is the number of participants in the third wave. It corresponds to 0.0067% of Spain's active population.

2.5 Estimand

The estimand of interest in this work is the matrix $\theta \in \mathbb{R}_{\geq 0}^{G \times G}$ with entries $\theta_{i,j}$, representing the mean number of contacts a person in age-group j has with persons in age-group i for all $G \in \mathbb{N}$ age-groups.

Figure 1 displays the mean number of total contacts between different age groups within a specific time period for the ground truth population. We can clearly see the intensity as well as intra- and intergenerational contact pattern, which both can have a direct influence on the spread of diseases through the underlying social network of the corresponding society.

[2] Distancia-Covid Survey, https://distancia-covid.csic.es/. I use data from the third wave (2020-12-14 to 2021-03-01).

2.6 Models

MRP consists of first, multilevel regression to produce small area estimates for poststratification cells and second, either poststratification, where the cell estimates are weighted according to the population cell size, or prediction, where the model response is predicted on the population.

We use the Bayesian MRP model by Palmer et al. [9], who estimate a population-level contacts distribution for the Spanish population based on data from the Distancia-Covid Survey. The multivariate response $y \in \mathbb{R}^G$ is the vector of mean number of contacts in G different age groups and the covariates include demographic and geographic characteristics of the survey respondents. Priors are set to the default priors of the R[3] package[4] that is used for the computation.

The multivariate model is defined as:

$$y_i \sim \mathcal{NB}(\mu_i, \omega) \tag{1}$$

$$\log(\mu_i) = \beta_0 + \sum_{c(i) \forall c=1,\ldots,C} u_{c(i)} \tag{2}$$

with default priors on the group-level residuals (group random effects), the overall mean, and the dispersion parameter (see Appendix A for detailed description).

The defined 58,240 poststratification cells of which approximately 27.76% are non-empty are based on 52 provinces, two gender categories, 14 age groups, four categories describing the three level of education, and ten occupation categories (detailed descriptions in Appendix A). The posterior distributions of the mean value for each cell are obtained by approximating it through 1000 draws.

The first model under investigation provides the population-level estimate by averaging the predictions after the poststratification step with respect to the poststratification cell sizes and takes the mean value over all draws as the final estimate - ignoring the reciprocal nature of contacts.

The second (equally weighted MRP) and third (sample weighted MRP) models are extensions that leverage the network information by enforcing reciprocity constraints. They take the estimated mean number of contacts and scale them on population level, resulting in a total number of contacts. The final estimate for mean number of contacts for a person in an age group j to persons in another age group i is the average of total contacts from j to i and total contacts from i to j, scaled to the inverse of the size of the age group j. The equally weighted reciprocity model uses the average with respect to the age group sizes on population level (see Eq. (3)) and the sample weighted uses the average with respect to weights according to the age group sizes in the samples and the population (see Eq. (4)).

$$\hat{c}_{i,j}^{\text{rec, ew}} := \frac{1}{2N_j}\left(\hat{c}_{i,j}N_j + \hat{c}_{j,i}N_i\right), \tag{3}$$

[3] R version 4.2.1 (2022-06-23) – "Funny-Looking Kid".

[4] We use the R package "brms", which is an interface for Stan.

where $\hat{c}_{i,j}$ are the estimated mean number of contacts from age group j to age group i, N_j the size age group j in the population.

The sample weighted extension applies following weighted average:

$$\hat{c}_{i,j}^{\text{rec, sw}} := \frac{1}{N_j}\Big(w_{i,j}\hat{c}_{i,j}N_j + w_{j,i}\hat{c}_{j,i}N_i\Big) \tag{4}$$

with weights

$$w_{i,j} := \begin{cases} \frac{n_j}{n_i+n_j} & \text{if } n_i + n_j > 0 \\ \frac{1}{2} & \text{otherwise,} \end{cases} \tag{5}$$

defined as fractions corresponding to the age groups sizes in the sample n_i.

2.7 Performance Measures

I quantify the methods performance by estimating a set of measures as well as their Monte Carlo standard errors (in order to capture the uncertainty originating from the simulations). I present an overview of the measure, the definition, the estimate, and the Monte Carlo SE in the Appendix A).

The percentage bias shows the average tendency of the estimated mean number of contacts to be larger or smaller than the true value. It therefore captures the accuracy of the applied methods in estimating the intensity of contacts between age groups and is of highest interest in this study. Furthermore, I estimate the empirical standard error that reflects the long-run standard deviation of the estimates over all conducted simulations and the Mean Squared Error as a measure of overall accuracy (the results are available in the Appendix A).

In addition to these performance measures, I include a measure that evaluates the performance in the context of networks. I am interested in the heterogeneity of contacts which is defined in this work through the diversity in number of contacts in the population, because this can have an influence on the spread of diseases through the underlying network. In the event of homogeneity, every person in the population would have exactly the same number of contacts. In case of heterogeneity there exists a diversity of different number of contacts.

To measure the heterogeneity in the network, I choose the normalized Heterogeneity Index by [5]. For a network with degree distribution $P(k)$ and N number of nodes, it is defined as $H_m := \frac{h}{h_{het}} \in [0, 1]$, where the non-negative result of the square root of $h^2 := \frac{1}{N}\sum_{k_{min}}^{k_{max}}(1 - P(k))^2, P(k) \neq 0$, is normalized with respect to its value for a completely heterogeneous network (all possible degrees exist): $h_{het} = 1 - \frac{3}{N} + \frac{N+2}{N^3}$.

The index therefore has a lower bound of 0, which corresponds to a completely homogeneous network, in which only for one $k_0 \in \{k_{min}, \ldots, k_{max}\}$ the value of $P(k) \neq 0$ is, hence $P(k_0) = 1$ and the sum vanishes.

3 Results

The first sampling scenario yields non-convergent transitions of the methods and shows that the models can not be applied in this extreme case.

3.1 Effect of Reciprocity - DCS

I apply the Bayesian MRP and its extensions on the DCS samples and obtain for each of the samples the mean number of total contacts for each age group pair. I use them together with the true value to estimate the percentage bias for each age group pair. The results are displayed in Fig. 2.

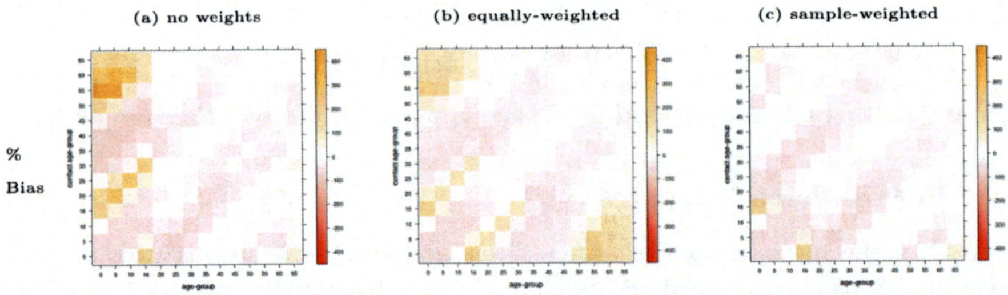

Fig. 2. Percentage bias of the estimators of the methods Bayesian MRP (a), and its extension with equally-weighted (b) and sample-weighted (c) reciprocity constraint, visualized as age-stratified matrices. The inclusion of equally weighted reciprocity yield to a "mirror" effect of the highly biased area visible in Fig. (a), which is avoided by including sample weighted reciprocity. The accuracy of the estimation in case of completely missing data (contacts within the under 20 years old group) are the same in all three cases, therefore there is no effect of including the reciprocity in this case.

First, the addition of the equally weighted reciprocity constraint introduces bias in an area where the estimations were previously fine: it "mirrors" and distributes the bias for the contacts from the underage age groups to all other age groups (compare Fig. 2 (a) left side, vertical bar) to the corresponding inverse contacts (compare bottom bar in Fig. 2 (a) and (b)). The reason for this is that the estimates that were obtained from small/zero-size data and therefore are highly biased obtain the same weight in the final result as the estimates that are based on richer data. Hence, the final result is perturbed by the first one. As a result, the inclusion of sample size information - and therefore the value trustworthiness of the estimates - by Bayesian MRP with sample-weighted reciprocity fixes this problem (see Fig. 2 (c)).

Moreover, the results show that the percentage bias of the estimators decreases by including reciprocity constraints, in particular in case of the sample-weighted version. I find that the estimates are higher by 83.17% on average in the contact estimations from children and adolescents to adults if no weights are applied, 33.71% for equally weighted and 15.76% lower in case of the sample weighted version. That means for a true value of mean number contacts 0.35, the unweighted MRP would overestimate the contacts with a value of 0.65 and the sample weighted version slightly underestimate it with a value of 0.29.

One area of high interest is the contacts among children and adolescences (within the age groups [0, 5), [5, 10), [10, 16), [16, 20)), because we are missing information on contacts in the sample completely. The estimated performance measure indicates that there is no effect of reciprocity on the accuracy of the estimates - in all three models the average percentage bias is approximately −3.85% and Fig. 2 shows that the contact pattern is not captured (underestimation where we expect higher intensity of contacts and underestimation where we expect lower intensity).

I further examine the contact pattern in more detail by partitioning the estimand into four areas of interest and calculating the average of the estimated percentage bias for these. As a result I can learn about the methods' ability to capture the specific pattern of contacts between different age groups.

The four parts are, first, the intragenerational contacts, which are visible as the diagonal in the age-stratified contact matrix[5], second, the upper parallel to the diagonal line as well as the lower parallel line, which both reflect intergenerational contact pattern[6], and fourth the remaining area besides those parallels and the diagonal.

Table 1 displays the results of the average estimates. The inclusion of reciprocity constraints improves the estimated number of intragenerational contacts slightly, but they are still underestimated (on average by 15.05%) and the intergenerational contacts are also more accurate after the inclusion of reciprocity and now are overestimated by 8.65% on average. Together with the result for in the remaining area in case of MRP with sample-weighted reciprocity (average underestimation by 10.61%), I can conclude that the method is not evening out all pattern, but instead capturing the contact structure of the different age groups.

3.2 Effect of Sampling Scenario - Sample Weighted MRP

I distinguish in this part two areas of interest. First, contacts among children and adolescences and second, contacts from and to this group as well as within adult contacts.

Independent of the sample scenario, the method performs poorly in capturing the intensity as well we the contact pattern for contacts within the younger age groups. The estimated mean number of contacts ranges from an average underestimation of 52.34% (CSRS) and 54.31% (DCS) to an average overestimation of approximately 97.14% (CSRS) and 82.38% (DCS). Figure 3 clearly shows this result in the bottom left corners.

In case of the CSRS scenario provides MRP with sample weighted reciprocity constraint excellent results (compare Fig. 3), in particular for estimating the

[5] For the diagonal, I use the age-group pairs starting from [0 ± 5, 0 ± 5) diagonally up to [60 ± 5, 65+ ± 5) as well as the neighboring group pairs.

[6] For the upper parallel line, I use the estimated mean number of contacts from age groups [0, 5) to age group [30, 45) diagonally up to the contacts from age group [35, 40) to group 65+. For the lower parallel line I use the contacts from age group [30, 35) to [0, 5) diagonally up to contacts from age group 65+ to [25, 35).

Table 1. Mean percentage bias of estimates of mean number of contacts within the same age group ±5 years (intra), from younger to older age groups with age difference between 30 and 45 years (top), from older to younger age groups (with same age differences), the mean independent of direction of intergenerational contacts (mean inter) and the mean percentage bias for the remaining age group pairs (non-parallel). Including reciprocity improves the ability to capture contact pattern between different age groups, in particular in case of intergenerational contacts (top, bottom).

Model	intra	top	bottom	mean inter	non-parallel
MRP	−16.12	−36.85	9.36	−13.74	24.7
MRP + ew	−16.12	−13.87	−13.26	−13.57	24.68
MRP + sw	−15.03	8.31	8.98	8.65	−16.02

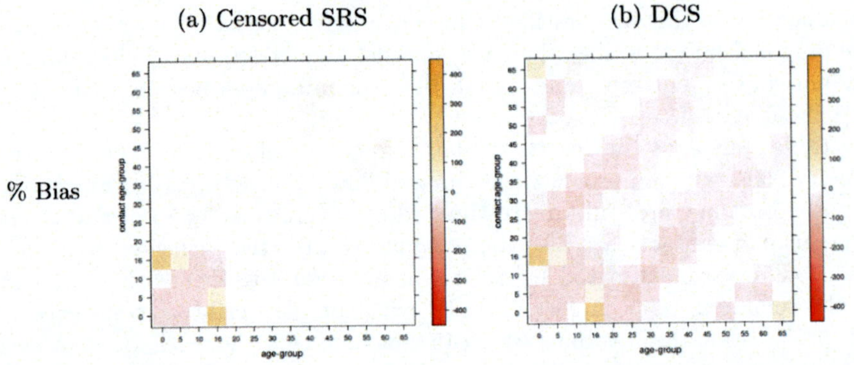

Fig. 3. Percentage bias of the estimators of the sample weighted Bayesian MRP for the CSRS scenario (a) and the DCS scenario (b). Independent of the sampling scenario fails the method to capture accurately the contact structure within younger age groups and tends to underestimate intragenerational contacts but over- and underestimate (CSRS ca. 7% and DCS ca. −11%) contacts in between generations.

mean number of contacts from children and adolescences to adults, for which only contact data in vice versa direction is available.

The effect of using highly selective samples (DCS sampling scenario) on the accuracy reflects in the overall mean percentage bias, which is only 4.31% for the CSRS and −11.82% for the DCS sampling scenario.

Distinguishing again between the intra-, and inter-generational pattern, I find that first, intragenerational contact pattern are for both scenarios underestimated by approximately 15% on average, second the intergenerational contact pattern are very accurate in the CSRS (-0.13% on average), but overestimated by 8.65% in the DCS scenario, and third, in the CSRS scenario the method tends to overestimate the contacts in the remaining area (by 7.25% on average), but underestimate them in the DCS case (by 10.61% on average).

3.3 Heterogeneity Index

Another focus in this work is the heterogeneity with respect to the diversity in node degrees as distinguished from heterogeneity associated with network structure.

The normalized Heterogeneity Index for the Labor Force Survey is $4.7 \cdot 10^{-4}$. In order to get an idea of the magnitude of this value, I created 50 random graphs that are known to be homogeneous (Erdős-Rényi model (ER), Watts-Strogatz model (WS)) and heterogeneous (Barabási-Albert model (BA)). I generate those random graphs with the same number of nodes and edges as the ground truth network[7].

The homogeneous models yield an average normalized Heterogeneity Index of $6.4 \cdot 10^{-4}$ (ER) and $5.7 \cdot 10^{-4}$ (WS) and the heterogeneous models yield an average Index of $70.1 \cdot 10^{-4}$. This means, that the ground truth network can be considered a homogeneous network according to the diversity in contacts.

The average index for the results of applying MRP for the CSRS samples is $5.5 \cdot 10^{-4}$ and $6.3 \cdot 10^{-4}$ for the DCS samples. Moreover, the variation of the index is higher in case of the biased DCS samples.

Hence, the simulation study shows that the homogeneity is slightly underestimated for both kind of samples and the normalized Heterogeneity Indices are similar to the corresponding homogeneous null models. Although the indexes are larger for our estimated networks the results can still be considered as homogeneity in both sampling scenarios. This can change if we examine the diversity associated to different age groups.

4 Limitations and Further Research

The main limitation of this work is that I could only conduct this study for in-home contacts, because no real representative ground truth data for other contacts is available for Spain at the moment. Therefore the results can not be transferred to other type of contacts than in-home contacts for which the assumption holds that individuals have contact with everybody that is also living in the same household.

Another limitation concerns the stratification. Here, as a first approach to the topic, I only consider age-stratified contacts, although not only ages of the individuals, but also their locations are related to the possibility and number of contacts [8]. A related limitation concerns the choice of age groups and their size. I assume that using exact ages could push to computational limits. The

[7] Due to the algorithms of the random graph generators is it not possible to fix the number of edges for the WS and the BA model. I create networks with 93,755,176 (WS), respective 93,755,173 (BA) edges, which are approximately 7.08% less than the number of edges in the ground truth network (100, 897, 435). The neighborhood within the nodes of the lattice are connected in the WA model is set to 2, which is approximately the ratio between number of nodes and edges. I decide a random rewiring with probability of 0.5.

Labour Force Survey provided 5-year age groups[8] and I decide to work with them, because they are the smallest grouping available.

Further limiting is the setup of the methods under investigation. The applied Bayesian MRP model that I use is only one of many possible configurations. Although the model already yield excellent results [9] it would be interesting to see if, first, group-level predictors can be used so that the model could better account for similarities between differently sized groups, and second, different priors can improve the estimation, in particular for the censored age groups. Besides expert knowledge on contacts between age groups, one could utilize structured priors as suggested by [3].

The approach of measuring to what extend the methods capture the inter-, intra-, and between generational contacts is only based on the visible parallels of the age-stratified contact matrix of the Labour Force Survey. This can be improved by utilizing or developing a network based measure for (age group) clustering.

The Heterogeneity Index by [5] can not be directly applied in combination with the reciprocity constraints and needs to be adjusted or even extended to an age-stratified Heterogeneity index, that can measure the diversity in degrees with respect to age groups.

5 Conclusion

In this paper, I have examined the ability of Bayesian MRP and its extensions that leverage characteristics of the underlying social network of Spain's population to estimate, based on highly selective and censored samples, the intensity as well as contact pattern within households.

I have considered three different sample scenarios that reflect real world highly biased and censored samples with the aim of examining the effect of change in bias.

I find that the inclusion of equally weighted reciprocity can introduce more bias and the sample weighted version improves the accuracy of the results. The intensity of contacts as well as inter-, intra- and in between-generational heterogeneity in contact patterns is well captured by sample weighted MRP, but still slightly over- and underestimated in the highly selective samples.

Nevertheless, all three methods fail to estimate neither the contact intensity nor the structure well within the younger age groups for which no information was included in the samples. Reciprocity inclusion therefore can only lead to improvements if at least some information is available.

The same result holds for the change in sampling scenario. For both, CSRS and DCS, the estimations are highly inaccurate for these missing age groups.

I further find that the effect of the increased bias in the samples (DCS) reflects in the overall accuracy of the sample weighted MRP. The method tends

[8] The age groups in the Labour Force Survey are not of equal size. The exception are the age groups [10, 16), [16, 20), and 65+ which contain 6, respectively 4, and all different ages above or equal 65.

underestimate the intragenerational contacts, independently of the scenario and the increase in bias in the samples lead to different over- and underestimation of generational contact pattern. In case of the DCS sample, the methods overestimates the intergenerational contacts, but underestimates all other. Hence, MRP with sample weighted reciprocity yield a slightly different structure than actually exists in the network.

In the final analysis of the network measure, I find that the biases in the samples lead to higher uncertainty regarding the diversity of estimated degrees in case of the Bayesian MRP. Independent of the sampling scenario I further find a slight overestimation in direction of heterogeneity. However, in comparison to typical homogeneous and heterogeneous null models (Erdős-Rényi, Watts-Strogatz, Barabási-Albert random graphs), I can conclude that the true as well as the estimated networks can be considered as homogeneous.

References

1. Breen, C., Mahmud, A., Feehan, D., et al.: Estimating subnational age-specific contact patterns using multilevel regression with poststratification. SocArXiv. November 10 (2021)
2. Feehan, D.M., Mahmud, A.S.: Quantifying population contact patterns in the united states during the covid-19 pandemic. Nat. Commun. **12**(1), 1–9 (2021)
3. Gao, Y., Kennedy, L., Simpson, D., Gelman, A.: Improving multilevel regression and poststratification with structured priors. Bayesian Anal. **16**(3) (2021). https://doi.org/10.1214/20-ba1223, http://dx.doi.org/10.1214/20-BA1223
4. Hamilton, D.T., Handcock, M.S., Morris, M.: Degree distributions in sexual networks: A framework for evaluating evidence. Sex. Transm. Dis. **35**(1), 30 (2008)
5. Jacob, R., Harikrishnan, K., Misra, R., Ambika, G.: Measure for degree heterogeneity in complex networks and its application to recurrence network analysis. Roy. Soc. Open Sci. **4**(1), 160757 (2017)
6. Klepac, P., et al.: Contacts in context: large-scale setting-specific social mixing matrices from the BBC pandemic project. MedRxiv (2020)
7. Morris, T.P., White, I.R., Crowther, M.J.: Using simulation studies to evaluate statistical methods. Statist. Med. **38**(11), 2074–2102 (2019). https://doi.org/10.1002/sim.8086, https://onlinelibrary.wiley.com/doi/abs/10.1002/sim.8086
8. Mossong, J., Hens, N., Jit, M., Beutels, P., Auranen, K., Mikolajczyk, R., Massari, M., Salmaso, S., Tomba, G.S., Wallinga, J., et al.: Social contacts and mixing patterns relevant to the spread of infectious diseases. PLoS Med. **5**(3), e74 (2008)
9. Palmer, J.R.B., Ottow, R., Bartumeus, F.: Households and person-to-person contacts in Spain's Covid-19 lockdown: Patterns of social mixing estimated from a web survey [unpublished work] (2021). working paper
10. Pawel, S., Kook, L., Reeve, K.: Pitfalls and potentials in simulation studies. arXiv preprint arXiv:2203.13076 (2022)
11. Wallinga, J., Teunis, P., Kretzschmar, M.: Using data on social contacts to estimate age-specific transmission parameters for respiratory-spread infectious agents. Am. J. Epidemiol. **164**(10), 936–944 (2006)

Academic Mobility as a Driver of Productivity: A Gender-centric Approach

Mariana Macedo[1(✉)], Ana Maria Jaramillo[2], and Ronaldo Menezes[2,3]

[1] Center for Collective Learning, ANITI, University of Toulouse, Toulouse, France
mmacedo@biocomplexlab.org

[2] Computer Science, University of Exeter, Exeter, UK
ajaramillo@biocomplexlab.org, r.menezes@exeter.ac.uk

[3] Computer Science, Federal University of Ceará, Fortaleza, Brazil

Abstract. pSTEM fields (Physical Sciences, Technology, Engineering and Mathematics) are known for showing a gender imbalance favouring men. This imbalance can be seen at several levels, including in university and industry, where men are the majority of the posts. Academic success is partly dependent on the value of the researchers' co-authorship networks. One of the ways to enrich one's network is through academic movement; the change of institutions in search of better opportunities within the same country or internationally. In this paper, we look at the data for one specific pSTEM field, Computer Science, and describe the productivity and co-authorship patterns that emerge as a function of academic mobility. We find that women and men both benefit from national and international mobility, women who never change affiliations over their career are rarely well-cited or highly productive, and women are not well-represented in the overall top-ranking researchers.

Keywords: Academic mobility · Gender inequality · Science of science · Network science · Data science

1 Introduction

The environment of academia is very competitive, and people are looking for opportunities to make their ideas more visible to a larger share of the community. Competitiveness arises from the mechanisms leading to promotion and tenure processes, which require academics to reach certain targets. Success in science, and hence one's career, is linked to productivity in terms of the number of publications and citations these works attract [31]. In this environment, and especially in pSTEM fields (Physical Sciences, Technology, Engineering, and Mathematics), gender plays an important, and sometimes discriminatory, role [6,29,33].

When becoming a top-ranked computer scientist, men's and women's career patterns differ significantly, as we have previously discussed [15]. We observed differences in network characteristics, such as the level of repeated co-authors, the composition of the co-authorship network, and network density, among others. However, one important dimension is neither considered in our previous

A. S. Teixeira et al. (Eds.): CompleNet 2023, SPCOM, pp. 120–131, 2023.
https://doi.org/10.1007/978-3-031-28276-8_11

work nor related to the prior literature on gender inequality: academic mobility (affiliation change) along career paths. Here, we use the same dataset as our previous work [15]. But, this study focuses on the role of academic mobility in the gender patterns of co-authorship networks' evolution, productivity, research impact, as well as the representation of women and men in top-ranked positions.

Academic mobility, understood as a change of affiliation, has frequently been shown as a mechanism to increase and diversify the co-authorship networks [4], with positive effects on career advancement due to fostering productivity [20] and citations [3,13,24]. However, there is evidence that academic mobility can lead to inequalities and biases when there are no research policies to properly integrate underrepresented groups framed by class [22], gender [19,25], race, ethnicity [21] and language [20].

Not all types of mobility are equal. Language barriers make certain movements more likely than others, but also other social constructs can influence decisions, including costs of migration [19], family ties [17], and the research field [3]. For example, people working in medical sciences, accounting, or law are likely to avoid international mobility because regulations vary significantly from country to country, forcing people to have their background revalidated or undergo extra training [12]. For this reason, we look here at two types of mobility: *(1)* international mobility, when a researcher changes their country of affiliation, which generally means a change in the place of work, and *(2)* national mobility, when a researcher change their place of work but not the country. This division will allow us to expand our work to other fields beyond Computer Science. International migration can have different effects on people's collaboration networks; it is expected that international migration will add more diversity to the researcher's co-authorship network and will lead to a broader view and recognition of their research. The patterns in the two groups mentioned above are compared to the ones of non-mobile researchers, those who never changed their affiliation.

According to our data, women and men can reach higher productivity levels in changing institutions (nationally or internationally). Nevertheless, men still consistently benefit more from the movement and women are better represented in top-ranking positions when we consider only non-mobile researchers. Surprisingly, national mobility presents the highest under-representation of women in the top 1% and 5% researchers than international mobility. Despite gender differences in productivity and women's representation, women and men tend to move around the same year of career length. In future, we intend to investigate whether these patterns are consistent with other research fields and datasets.

2 Data

Computer Science is a field in which women have been a minority over many years [6,29,33]. Countries have adopted new policies to encourage women to enter and not drop out of the field [7,10,28], but we still do not have women well-represented in top-ranked positions [32]. In this paper, we study the gender role of academic mobility from the data of the Computer Science field of the ACM

Table 1. Number of authors and papers per gender and category. There is an overlap in the number of papers published by `women` and `men` for the papers with co-authors from both genders.

Quantity	Category	`women`	`men`	Total
	`all`	14,433 (16%)	77,344 (84%)	91,777
Authors	`international`	4,859 (14%)	29,254 (86%)	34,113
	`national`	3,235 (15%)	17,897 (85%)	21,132
	`non-movers`	6,339 (17%)	30,193 (83%)	36,532
	`all`	176,325	746,211	809,397
Papers	`international`	96,571	491,884	544,698
	`national`	56,730	288,231	324,971
	`non-movers`	37,507	170,581	198,330

(Association for Computing and Machinery) Digital Library. The data were collected by Divakarmurthy and Menezes [9], and the gender of the authors was inferred by Jaramillo et al. [15] using Genderize [1] and Namepedia [2]. The data comprises bibliometric information for research papers published between 1980 and 2012 (e.g. authors, authors' affiliation, title, keywords, and year). We use the authors' affiliations from the publications to trace the researchers' mobility. An institution's name can be written differently, for example, *University of Toulouse* and *Université de Toulouse* refer to the same institution. Faustino et al. [11] pre-processed the data to match these affiliations, and we reuse their approach here.

Based on the affiliations provided by each researcher in each publication, we classify researchers according to their movement as *(i)* non-movers (`non-movers`), researchers with the same affiliation in all the publications; *(ii)* national (`national`), researchers with all the affiliations from the same country, but at least 2 different affiliations; and *(iii)* international (`international`), researchers with at least 2 affiliations from different countries. In total, we have 14,433 women and 77,344 men in our data, with a women's proportion of around 16%, which is slightly consistent across the categories of career movements (see Table 1). For the case of publications, at least one woman participates in around 22% of the papers in our dataset. For each researcher, we consider the first year of publication available in our dataset as year 0 of their career length. In the shorter career length, there are more authors and papers as researchers have different career lengths and the longer the length, the fewer people belong to that group (see Fig. 1). We analyse the first 25 years of their career, as we can have at least 300 researchers for any category.

3 Methods and Results

3.1 Mobility and Co-authorship Networks

Co-authorship networks have been related to increments in researchers' productivity [5,14,15,23,31]. It is, however, still necessary to study how academic

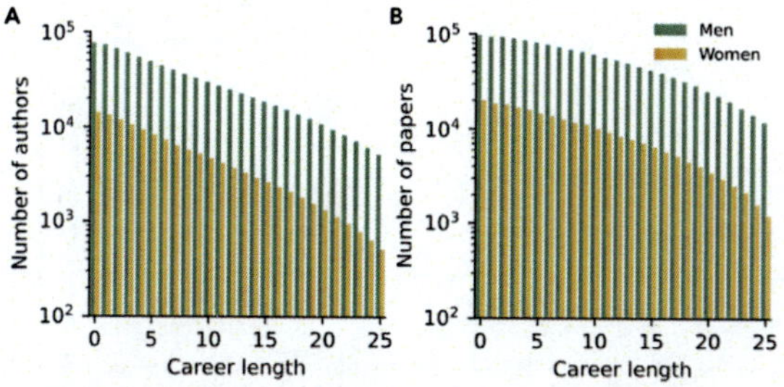

Fig. 1. **(A)** Number of authors per career length. **(B)** Number of papers in the dataset per career length.

mobility can shape the co-authorship networks of women and men in an ego-centric way. Here, we analyse the gender differences in the temporal patterns of the co-authorship networks between the researchers across career movement categories. For each researcher, we constructed ego co-authorship networks over the career length ℓ. The authors i and j connect with a link weighted by the number of co-authored papers published in our dataset, $W_\ell(i,j)$. In Fig. 2, we show the temporal evolution of 12 metrics of the ego co-authorship networks for each career length that we explained in our previous work [15].

We show topological characteristics in the first row of Fig. 2. Both women and men who are `national` and `international` movers have more co-authors than `non-movers` (Fig. 2A). Previous literature has shown that living abroad can increase the social capital gained by proximity; if there is no discrimination, social ties could expand and diversify [30]. Consequently, the high number of co-authors translates into a higher weighted degree (Fig. 2B) (more co-authored papers with the same researchers), a lower density (Fig. 2C) (as an effect of the increased network) and lower average clustering (Fig. 2D) (multiple co-authors working in separate groups). On the second row of Fig. 2, we show the composition of corrected proportions of both total and new co-authors from each gender as alters. As in our previous paper [15], each gender is more likely to collaborate with the same gender, correcting by the gender distribution in the dataset (Fig. 2E and Fig. 2F). Homophily does not change with academic mobility, and women `national` movers are the group with the highest homophily. On the last row of the Fig. 2, we show the triadic closure and its maintenance when each gender is an alter connecting the ego node with a new node. Women are more likely to introduce/maintain co-authors to/with women, and the same happens to men (Fig. 2I–L). However, values smaller than 0.5 in Fig. 2I and Fig. 2J show that researchers are more likely to have completely new co-authors.

In summary, the change of affiliations and countries has a small effect on the characteristics of egos' co-authorship networks. We found that academic mobility has a positive effect on expanding the number of co-authors, which resembles the ones expected from top-ranking researchers [15]. Nevertheless, career movements

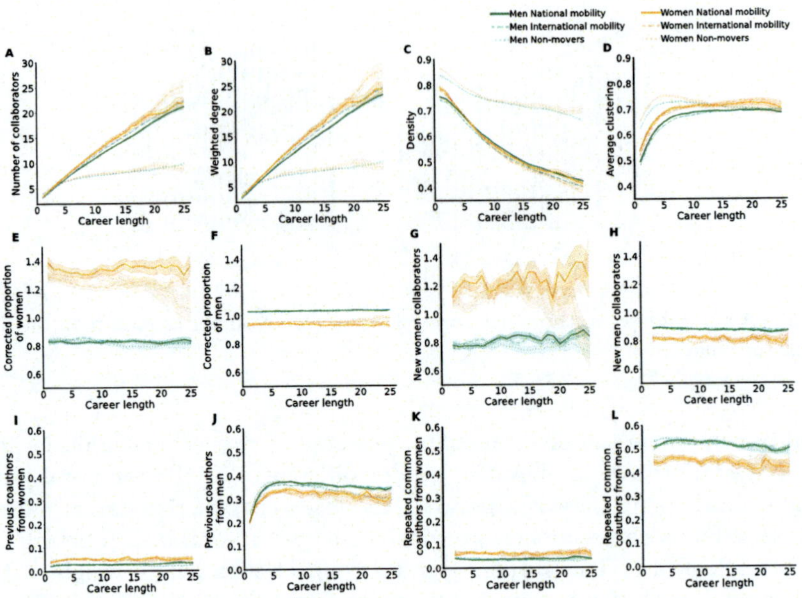

Fig. 2. Co-authorship network analysis per gender and category. **(A)** Number of co-authors **(B)** Weighted Degree **(C)** Density **(D)** Average clustering **(E)** Corrected proportion of `women` **(F)** Corrected proportion of men **(G)** New `women` co-authors **(H)** New `men` co-authors **(I)** Previous coauthors from `women` **(J)** Previous coauthors from `men` **(K)** Repeated common coauthors from `women` **(L)** Repeated common coauthors from `men`.

are beneficial to both women and men when it comes to the number of co-authors, potentially having a positive impact on productivity and citations.

3.2 Mobility and Productivity

Academic mobility positively impacts the productivity of researchers [3,13,20,24]. In this paper, we look at "productivity" as the number of published papers and the number of citations to measure the impact of researchers in Computer Science. Figure 3 shows that each gender increases their productivity over their career length regardless of the categories of movements. `International` movers benefit the most, followed by similar trends for `national` movers, with an average of 3 times higher productivity than `non-movers`.

In the analysis of the gender differences in Fig. 3B–D, we observe statistically significant differences in most career years. In the early career stages, women are slightly more productive than men, but after 10 years, men tend to keep up and increase their productivity faster. The highest differences between genders are for `national` movers. We argue that this difference might be due to the fact that

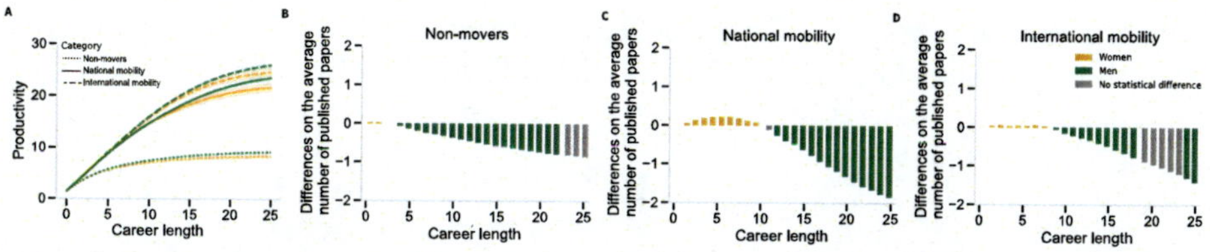

Fig. 3. Productivity over career year. **(A)** Cumulative number of published papers. Gender differences between the average productivity of **(B)** **`non-movers`**, **(C)** **`national`** movers and **(D)** **`international`** movers.

most movements occur within high-income countries, where more institutions have high international rankings.

Then, we study the gender differences when combining the number of national and international movements. In Fig. 4, our results indicate a higher fraction of women in the `non-movers` category (Position 1,1 on the heatmap) but a higher proportion of `men` on the rest of the heatmap. When we look at the productivity trends combining movement categories, Fig. 4B, we found that `men` tend to have higher productivity for `non-movers` and lower values of `national` and `international` movements. In contrast, women who moved three times have higher productivity than men in the same situation.

We analyse the relationship between productivity and citations in Fig. 6; the distribution of `women/men` and their fraction in the four quadrants of the plots. The smallest fractions for both genders are for `non-movers` (high-right quadrant: 0.16% `women` and 0.23% `men`), and the largest fraction of both genders are also for `non-movers` (low-left quadrant: 98.52% `women` and 97.85% `men`). The highest difference between the movement categories is for researchers in the quadrant of high productivity-low citations, with national and international movers having, on average, 10 and 8 times more than `non-movers`. Regarding citations, `women` in both quadrants of high and low productivity get no differences when moving nationally (3.35%) or internationally (3.34%). In contrast, the fraction of `men` slightly increases when moving internationally (5.02%) compared to nationally(4.3%).

Then, we use Gini coefficient [8], as an inequality metric, to measure the evenness from the distribution of productivity and citations (Table 2). We observe that `women` have a more evenly distributed number of papers and citations than `men`. The Gini coefficients are higher when considering the number of citations than the productivity, indicating that citations have higher variability. The gender differences from the Gini coefficients are higher for `national` movers. In contrast, `non-movers` researchers were the ones with the smallest Gini in productivity, and the highest in the number of citations, suggesting that their productivity does not translate literally to citations. The last aligns with Fig. 6 in the legend of the top panels, where national and international movers have higher correlation values ($S_M, S_W \approx 0.7$) than `non-movers` ($S_M, S_W \approx 0.5$).

International movers have Gini Coefficients more similar between productivity and citations, showing the smallest gender differences (Table 2).

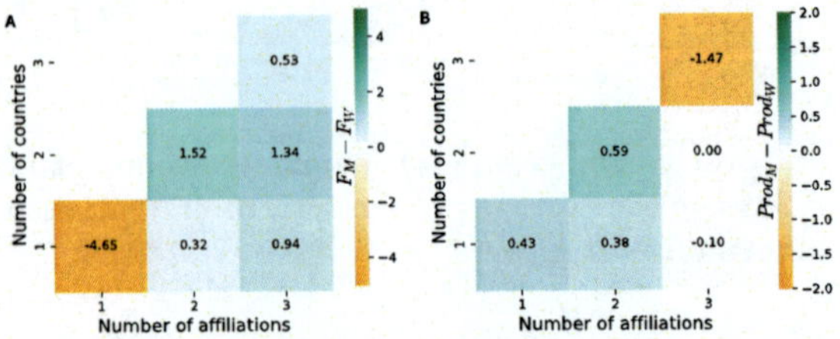

Fig. 4. **(A)** Differences on the fraction of `men` (F_M) and `women` (F_W) considering the number of countries and affiliations. **(B)** Differences on the productivity of `men` (Prod_M) and `women` (Prod_W). We do not consider a higher number of countries and affiliations, as the percentages of researchers are smaller than 1%.

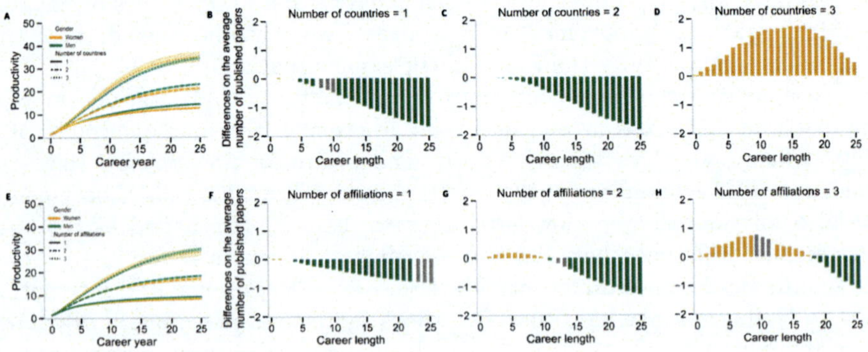

Fig. 5. **(A)** Productivity per gender and number of countries over the career length. Gender differences between the average productivity of researchers who worked in **(B)** 1 country, **(C)** 2 countries and **(D)** 3 countries. **(E)** Productivity per gender and number of affiliations over the career length. Gender differences between the average productivity of researchers who worked in **(F)** 1 affiliation, **(G)** 2 affiliations and **(H)** 3 affiliations.

In summary, `non-movers` have lower productivity and research impact from citations, not reaching the same levels of researchers who move nationally and internationally. The fraction of women and men decreases as we consider the number of countries and affiliations where researchers worked. Productivity correlates more with citations for movers, and researchers with international mobility are more productive and cited. When looking at national movers with high

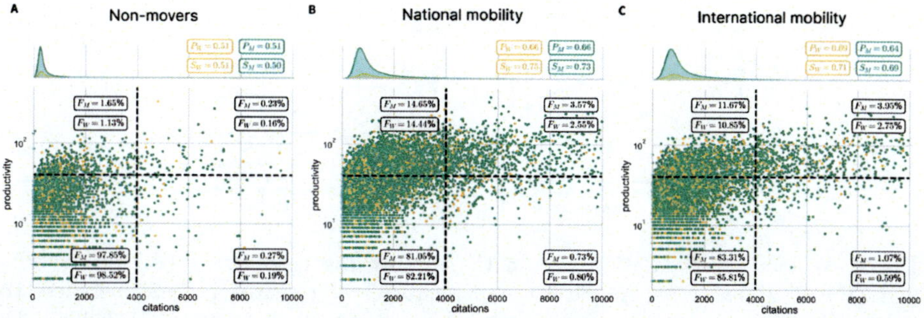

Fig. 6. Productivity versus citations across career movements: **(A)** Non-movers **(B)** National mobility **(C)** International mobility. Mobility has a role in how distributed `women` (yellow) and `men` (green) are in the plot, making the kurtosis smaller and increasing the number of productive and highly-cited researchers. The plot indicates the fraction of `women` (F_W) and `men` (F_M) for each quadrant, and it shows the Pearson ($P_{M|W}$) and Spearman correlations ($S_{M|W}$) between the productivity and citations for each gender.

productivity and high citations, there is a slightly higher fraction of women than men ($0.80 - 0.73 = 0.07\%$).

3.3 Mobility and Gender Differences

For Computer Science, women are generally underrepresented in the top-ranking positions [14,15,18]. Here, we test the hypothesis that `women` are even more underrepresented when considering career movements. Women's representation in the top-ranking decreases as we increase the percentage of people in the ranking related to productivity (Fig. 7A) and citations (Fig. 7B). Considering productivity, for the top 1% researchers, we see that `women` are much more underrepresented for the group of researchers who moved affiliations within the same country. In comparison, considering citations, the top 1% have much lower values of women representation for all the categories, and the top 1–10% for `national` and `international` movers are much similar. In general, we see that `women` are

Table 2. Gini coefficients from the productivity and citations' distributions.

Gini coefficient	Category	`women`	`men`	`(women-men)`
Productivity	`international`	0.5352	0.5352	0.0000
	`national`	0.5270	0.5403	−0.0133
	`non-movers`	0.4860	0.4945	−0.0085
Citations	`international`	0.6921	0.7044	−0.0123
	`national`	0.6966	0.7160	−0.0194
	`non-movers`	0.7191	0.7251	−0.0060

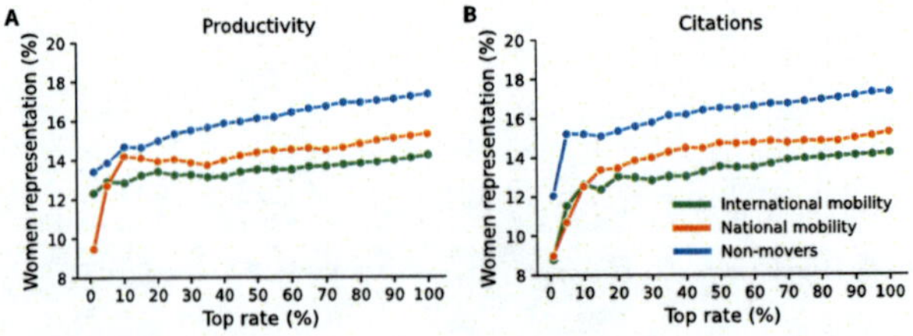

Fig. 7. Women representation (%) in the top-ranking considering **(A)** productivity and **(B)** citations as we increase the percentages of researchers. For instance, the top 100% researchers are the entire data equal to gender distribution in Table 1. For national mobility, the top 1% researchers reach the lowest women representation, equal to 9.48%.

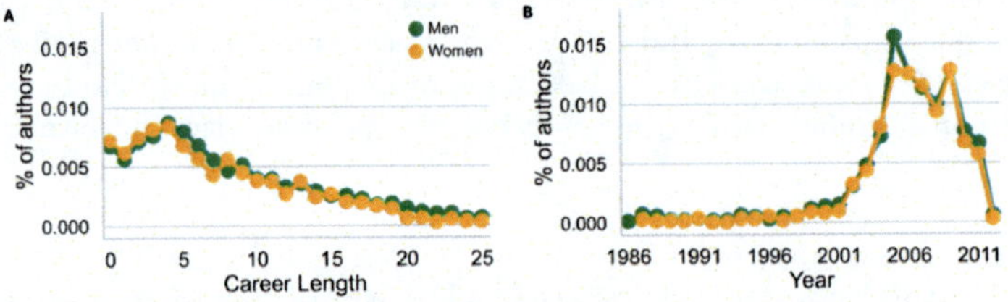

Fig. 8. Percentage (%) of women and men per **(A)** career length and **(B)** year.

more underrepresented in the top 1–15% across all categories and the percentages higher than 30% in the top-ranking slightly stabilize the women rate. We then test whether the career movements are happening at different rates for both genders over the career. In Fig. 8, we observe that the gender differences are small and that the patterns are similar across genders. Therefore, we argue that, regardless of gender differences identified in the number of co-authors and productivity, there are no indicators that when the movement occurs, it has a major role in the career. This is why researchers are more likely to move at the beginning of their careers.

In conclusion, women are not well-represented in the top-ranking, and researchers with national mobility showed the smallest representation of women for the top 1%. At the beginning of their careers, women and men are more likely to change affiliations, and there are no statistical gender differences in which year or career length they decide to move.

4 Discussion

Similar to market trends, scientific knowledge spreads through networks hopefully leading to impact [16,27]. Academic mobility, in which researchers change institutions across different cities and countries, has been studied by geographers and bibliometricians to understand the impact of the movements in researchers' careers and the spreading of knowledge [20,24,26]. When appropriate integration policies are in place, the researcher's productivity, co-authorship networks, and citations are strengthened. However, when the integration policies do not consider the external social constructs groups of researchers, the same strategies to pursue "successful" careers can be detrimental to certain groups because groups are not homogeneous [4]. For example, caregivers of children or relatives move less because they have insufficient financial or social support [25]. The sad reality is that women experience less mentoring support and more conflict when it comes to balancing family and academic responsibilities than men. In fact, a higher percentage of senior women researchers do not have children compared to senior men researchers [17]. Therefore, women and men in senior positions may have to make different decisions in their personal lives in order to maintain similar productivity levels.

In this paper, we examined the patterns in the career of women and men researchers in the dataset of publications from the ACM Digital Library (between 1980 and 2012). Using network analyses, we found similar characteristics across genders and career movements on the co-authorship networks. Nonetheless, we found differences in the number of co-authors that men and women gain over their careers, suggesting that changing affiliations nationally and internationally benefits productivity. Social ties can positively impact productivity, as writing papers collaboratively can speed up the process and lead to better quality work. However, we find that the small differences between the number of co-authors for women do not impact their productivity, making them more productive than men. Furthermore, as men are the majority in our data, gender homophily benefits high productivity levels more for men than for women.

We also found that the gender differences in productivity between `non-movers` researchers are smaller than when compared to the movers. Perhaps it is a case of the rich-getting-richer or selection bias, which could make men more likely to be hired in high-ranked institutions than women. The gender gap in women's representation within high-ranked institutions within and across countries needs to urgently be investigated. For instance, how is the relationship between moving from a developing nation to a developed nation different from moving across developed nations from a gender-centric perspective?

It is important to note that our analyses are based on assumptions and definitions limited by the available data. In our analyses, we first assume that gender is binary, and therefore no physical or biological differences between people play a role. Secondly, we did not analyse the authors with unisex names, and the libraries used to detect names still need to be more effective for Asian names. We attempted to overcome this limitation by using databases of gendered names, but we did not ensure that gendered detection was unbiased. We also highlight that women are still underrepresented in senior career stages, which can be caused by a lack of data. Nevertheless, we cannot guarantee that this unbiased disproportion is not affected by a lack of data. Another limitation of tracing academic mobility from publications is that we should not only consider researchers who kept publishing in ACM venues. Since our data is limited to one dataset, it is possible that researchers changed their preferred form of publication. Given that computer scientists consider the ACM to be well respected and therefore important to their careers, we believe this is the exception and not the rule.

5 Conclusion

This paper demonstrates how gender and mobility may affect one's productivity and impact in computer science, as measured by papers and citations. Although there is no indication of differences when women and men change affiliations, gender differences in productivity increase over the career length. Indeed, changing affiliations benefit both genders, but there is still an open question of how women can make up for the advantages men have in their careers. Moreover, the policies have not yet translated into a decrease in the gender differences in productivity and citations for our data until 2011. Still, the percentage of women and men who can change affiliations has stayed the same over the years. One follow-up question is whether these changes are evenly distributed when compared to top-ranking institutions and others.

References

1. Genderize. https://genderize.io/. Accessed 11 Nov 2020
2. Namepedia. http://www.namepedia.org/. Accessed 11 Nov 2020
3. Aman, V.: A new bibliometric approach to measure knowledge transfer of internationally mobile scientists. Scientometrics **117**(1), 227–247 (2018). https://doi.org/10.1007/s11192-018-2864-x
4. Barjak, F., Robinson, S.: International collaboration, mobility and team diversity in the life sciences: impact on research performance. Soc. Geography (2008). https://doi.org/10.5194/sg-3-23-2008
5. Bravo-Hermsdorff, G., et al.: Gender and collaboration patterns in a temporal scientific authorship network. Appl. Network Sci. **4**(1), 1–17 (2019). https://doi.org/10.1007/s41109-019-0214-4
6. Casad, B.J., et al.: Gender inequality in academia: problems and solutions for women faculty in stem. J. Neurosci. Res. **99**(1), 13–23 (2021)
7. Cech, E.A., Blair-Loy, M.: The changing career trajectories of new parents in stem. Proc. Natl. Acad. Sci. **116**(10), 4182–4187 (2019)
8. Ceriani, L., Verme, P.: The origins of the Gini index: extracts from Variabilità e Mutabilità (1912) by Corrado Gini. J. Econ. Inequal. (2012). https://doi.org/10.1007/s10888-011-9188-x
9. Divakarmurthy, P., Menezes, R.: Area diversity in computer science collaborations, pp. 2041–2042, Trento Italy (2012). https://doi.org/10.1145/2245276.2232115
10. Dryburgh, H.: Underrepresentation of girls and women in computer science: classification of 1990s research. J. Educ. Comput. Res. **23**(2), 181–202 (2000)
11. Faustino, J., Iyer, N., Mendonza, J., Menezes, R.: Characterizing the dynamics of academic affiliations: a network science approach. In: Barbosa, H., Gomez-Gardenes, J., Gonçalves, B., Mangioni, G., Menezes, R., Oliveira, M. (eds.) Complex Networks XI. SPC, pp. 393–404. Springer, Cham (2020). https://doi.org/10.1007/978-3-030-40943-2_33
12. Glinos, I., Wismar, M., Buchan, J.: Governing health professional mobility in a changing Europe. Eur. J. Pub. Health (2014). https://doi.org/10.1093/eurpub/cku164.121
13. Horta, H.: Deepening our understanding of academic inbreeding effects on research information exchange and scientific output: New insights for academic based research. High. Educ. (2013). https://doi.org/10.1007/s10734-012-9559-7

14. Jadidi, M., Karimi, F., Lietz, H., Wagner, C.: Gender disparities in science? dropout, productivity, collaborations and success of male and female computer scientists. Adv. Complex Syst. (2018). https://doi.org/10.1142/S0219525917500114
15. Jaramillo, A.M., Macedo, M., Menezes, R.: Reaching to the top: the gender effect in highly-ranked academics in computer science. In: Advances in Complex Systems, vol. 24 (2021). https://doi.org/10.1142/S0219525921500089
16. Jepsen, D.M., et al.: International academic careers: personal reflections. Int. J. Hum. Resour. Manag. (2014). https://doi.org/10.1080/09585192.2013.870307
17. Leemann, R.J.: Gender inequalities in transnational academic mobility and the ideal type of academic entrepreneur. Discourse (2010). https://doi.org/10.1080/01596306.2010.516942
18. Lerman, K., Yu, Y., Morstatter, F., Pujara, J.: Gendered citation patterns among the scientific elite. Proc. Natl. Acad. Sci. **119**(40), e2206070119 (2022)
19. Leung, M.W.: Unsettling the Yin-Yang harmony: an analysis of gender inequalities in academic mobility among Chinese scholars. Asian Pac. Migr. J. (2014). https://doi.org/10.1177/011719681402300202
20. Marginson, S.: What drives global science? The four competing narratives. Stud. High. Educ. (2022). https://doi.org/10.1080/03075079.2021.1942822
21. Marmolejo-Leyva, R., Perez-Angon, M.A., Russell, J.M.: Mobility and international collaboration: case of the Mexican scientific diaspora. PLoS ONE (2015). https://doi.org/10.1371/journal.pone.0126720
22. McMahon, M.E.: Higher education in a world market. High. Educ. (1992). https://doi.org/10.1007/bf00137243
23. Newman, M.E.: The structure of scientific collaboration networks. Proc. Natl. Acad. Sci. **98**(2), 404–409 (2001)
24. Petersen, A.M.: Multiscale impact of researcher mobility. J. R. Soc. Interface (2018). https://doi.org/10.1098/rsif.2018.0580
25. Sautier, M.: Move or perish? Sticky mobilities in the Swiss academic context. High. Educ. **82**(4), 799–822 (2021). https://doi.org/10.1007/s10734-021-00722-7
26. Scellato, G., Franzoni, C., Stephan, P.: Migrant scientists and international networks. Res. Policy **44**(1), 108–120 (2015)
27. Shen, W., Xu, X., Wang, X.: Reconceptualising international academic mobility in the global knowledge system: towards a new research agenda. Higher Educ. **84**(6), 1317–1342 (2022). https://doi.org/10.1007/s10734-022-00931-8
28. Stephenson, C., et al.: Retention in computer science undergraduate programs in the us: Data challenges and promising interventions. ACM (2018)
29. Stout, J.G., Grunberg, V.A., Ito, T.A.: Gender roles and stereotypes about science careers help explain women and men's science pursuits. Sex Roles **75**(9), 490–499 (2016)
30. Urry, J.: Mobility and proximity (2002). https://doi.org/10.1177/0038038502036002002
31. Wang, D., Barabási, A.L.: The Science of Science. Cambridge University Press, Cambridge (2021)
32. Wang, L.L., Stanovsky, G., Weihs, L., Etzioni, O.: Gender trends in computer science authorship. Commun. ACM **64**(3), 78–84 (2021)
33. Xu, Y.J.: Gender disparity in stem disciplines: a study of faculty attrition and turnover intentions. Res. High. Educ. **49**(7), 607–624 (2008)

Check for updates

Getting the Boot? Predicting the Dismissal of Managers in Football

Mounir Attié(✉), Diogo Pacheco, and Marcos Oliveira

Computer Science, University of Exeter, Exeter, UK
{maa250,d.pacheco,m.a.oliveira}@exeter.ac.uk

Abstract. Football club managers have a challenging and remarkably volatile job—the practice of sacking and replacing managers is widespread in the modern game. However, it is still unclear what exactly motivates managerial dismissal in clubs. More than ever, high-quality statistics are available to clubs, suggesting that dismissal decisions tend to be well informed. Likewise, supporters on social media might also influence clubs' decisions. Here we propose machine learning models to characterize the determinants of managerial dismissals. Is social media pressure associated with managerial sacking? Yes! We fit multiple ElasticNet regularised logistic regression models using features based on the social pressure of fans on Twitter and football statistics, showing that our best model obtains a balanced prediction accuracy of 0.75.

Keywords: Football dismissal · Impact performance and success prediction · Machine learning prediction · Social media

1 Introduction

Football is one of the most beloved sport worldwide. It unites people across borders, builds bridges, and provides an escape for so many from the hardships of life. The sport, however, can also be the creator of tribulations, notably for the manager, or first-team coach, due to the volatile nature of their job. Managers come and go way quicker than anybody else in football, but the reasons for this phenomenon are yet to be discovered [2,5,21]. With ever-increasing popularity, football clubs have become massive commercial entities with revenues in the millions [12,13]. The result of this is that top football clubs are run in a similar way to top businesses—with a certain ruthlessness towards underperforming at every level [17].

Yet most football managers have very little time to achieve their goals. The average tenure of a manager in England's top division (the Premier League) was less than 22 months in 2015, and in the second division (the Championship), just around 10 months [20]. Undeniably, for any business, constant failure to achieve company objectives is detrimental to growth. However, the frequency of sackings may lead us to believe that there may be a certain unfairness. Rowe et al. [18] call this the ritual scapegoating theory, in which succession at any given sports team

A. S. Teixeira et al. (Eds.): CompleNet 2023, SPCOM, pp. 132–140, 2023.
https://doi.org/10.1007/978-3-031-28276-8_12

is inevitable despite many times there being no practical reason for it. Nonetheless, decision-makers at football clubs are thought to make well-informed decisions based on state-of-the-art statistics to decide whether to sack managers or not. Conversely, every club is run differently and follows different processes to determine the fate of a manager. These factors make this area of research complex and active.

Managerial turnover has been an area of research for many academics in sports (not only in football), attempting to understand the causes and consequences of managerial turnover for decades [1–4, 6–10, 23]. Several researchers have attempted to understand why a manager gets sacked, and if it is the consequence of solely losing a few games. The general consensus is that it isn't the case, but the determinants of dismissal can vary a lot. They often conclude there is missing data in their models, such as fan or media pressure. Public relations are critical in football, and it is almost impossible to neglect the impact that the fans have on players, directors, managers and on the club as a whole. Football fans are known to be prominent users on social media to vent and express their views on different footballing topics and issues [16]. Twitter, for instance, has been used to characterise the world cup [14] and to unveil rivalries [15]. More specifically, the sentiment of tweets have been used to predict football scores [11, 19]. Social media data has been used to justify general dismissals based on individual behaviour and privacy breaches [22], but to the best of our knowledge, it has never been used to predict sacking football managers or dismissals based on public opinion.

Previous works predicting dismissals mainly focused on probit and logit models, and survival models such as parametric proportional hazard. Unsurprisingly, they agree on the *team's win ratio* as an important determinant of a turnover. For instance, in NBA (basketball), the *team's win ratio* was the only significant determinant [9], but in NFL (American football) *player talent*, *coaching talent*, *on-field performance*, and *resource quality* were all important determinants [1]. Despite the lack of consensus, they offer valuable insights into other significant variables in their models. Forrest et al.'s [7] equally trained a probit model with football data from the Spanish first division and obtained quite different results. They concluded that features such as the *time of the season*, the result of the *last match*, a *previous dismissal* in the season, and a *relegation battle* are all determinants. These differences suggest singularities between different sports when sacking managers. Bachan et al. [3] found the *attendance at stadiums*, the *manager "internationalisation"*, and a *relegation battle* as relevant determinants of a dismissal. d'Addona et al. [8] concluded that *age* and *prior experience* of the manager, *recent* match *results*, and a *change* in league *position* as important factors to dismissal. Both works [3, 8] exploit time-discrete logit models, enabling time-based analysis and modelling. Chase et al. [6] also use a logit model, but not very successful, only predicting 2 out of the 7 managerial dismissals correctly with NFL data. Frick et al. [10] garnered data from Germany's first division (the Bundesliga) and built a mixed logit model. Despite achieving good results, their model is very complex and relies on individual characteristics rather than statistics.

Another popular way to predict managerial turnover is with a parametric hazard model. A survival model analyses the duration of time until the event "a manager is sacked" occurs. Hence, utilising continuous time periods instead of discrete ones. This can be an issue when there is missing data. Many studies use variations of this model, such as Cox proportional model and the Weibull model, whose variation stem from different mathematical assumptions [2,4,8,23]. They draw similar conclusions to the researchers utilising probit and logit, but have also found other determinants. For instance, Audas et al. [2] conclude that results in recent games are more relevant than those from older games, but also consider the position of the team in the table before and after the manager was hired a crucial factor. Van Ours et al. [23] found the result of the last four games as the most important determinant for dismissal. The second most important factor was the "cumulative surprise indicator" – calculated by comparing actual results and expected results based on bookmaker's odds.

This work uses machine learning models to find the determinants of a managerial dismissal. We use traditional football statistics and analyse emotions in tweets. The results reveal that the sentiment of fans combined with match stats from previous ten games can accurately predict dismissals.

2 Results

To model and predict the relevant determinants of managerial turnover, we collected statistics from England's top two divisions: the Premier League and the Championship, including over 100 features about managers and teams. We also used sentiment analysis to gather the sentiment of the fans on Twitter about their club and about each manager individually for 4 different seasons ranging from 2014/15 to 2021/22. We use this data to understand the determinants of a managerial casualty.

2.1 Models

We construct models that categorizes the outcome of a managerial tenure (i.e., sacked or not sacked) based on different sets of features, allowing us to understand the determinants of dismissals. In all models, we used a regularized logistic regression model (i.e., ElasticNet), allowing us to select features for the models efficiently. We build 7 models of increasing complexity, as described in Table 1, which have the following feature categories:

Manager's career This category evaluates the manager's reputation as a manager, based on their previous and current tenures, including the record (i.e., wins, draws, losses, points, PPM) achieved for different periods.

Manager's playing career This category evaluates managers' reputation during their playing career, based principally on the leagues they played in and the games played in those leagues.

Table 1. Models with increased levels of complexity.

Model	Description	# Features
Naïve model #1	Points per match (PPM) of last 5 matches	1
Naïve model #2	Points per match (PPM) of last 10 matches	1
Group model #1	Tweets' sentiment scores only	8
Group Model #2	Manager's career and playing career	31
Complex Model #1	Match stats and results of last 10 matches	28
Complex Model #2	Match stats, results of last 10 matches, and manager's record and career	59
Complex Model #3	Match Stats, results of last 10 matches, manager's record and career, and sentiment scores	67

Match stats This category is the broadest, evaluating all the match statistics for a certain number of games. These include possession, passing accuracy, shot accuracy, touches and 22 other analytics from a game.

Sentiment scores The positive and negative sentiment about the manager and club as measured using Vader and TextBlob.

We note that ElasticNet combines both $L1$ and $L2$ regularizations with wheighing controlled by an α value. To find the best α, we use k-fold cross validation with $k = 10$. We also use stratified sampling to ensure that the rate of sacked and not sacked manager are the same in both the training set and test set.

2.2 Model Comparisons

First, we compare the naive models, finding that the model fitted with the PPM for 10 games outperforms the PPM for 5 games across different metrics (see Table 2). This finding supports our decision of using statistics of the last 10 games in all the other models. Next, we examine the group models; one fitted with uniquely sentiment score (Group Model #1) and the other using managers reputation as a player and manager (Group Model #2). These two models are based on the assumption that the sacking of a manager is based solely on the

Table 2. Performance metrics for all models.

Model	Precision	Recall	F1 score	Accuracy
Naïve Model #1	0.66	0.80	0.72	0.70
Naïve Model #2	0.79	0.84	0.82	0.82
Group Model #1	0.63	0.62	0.63	0.63
Group Model #2	0.77	0.83	0.80	0.79
Complex Model #1	0.82	0.92	0.87	0.84
Complex Model #2	0.86	0.91	0.89	0.89
Complex Model #3	0.82	0.96	0.88	0.85

sentiment of the fans and solely on the manager's past, respectively. We find that Group Model #2 outperforms Group Model #1 based on the AUC (0.88 and 0.67 respectively). When examining the complex models, we find that they perform better than these previous models, as shown in Table 2.

These models are all viable, but these analyses thus far have focused solely on the performance of the training set. We are interested in the generalizability of the model. For that purpose, we analyze the performance of these models in the test set as well. We use unweighted average recall (UAR) as the primary metric to evaluate the models' ability to predict outcomes, as it takes into account unbalance in the data. In this analysis, we focus on the Complex models and the Naive Model #2, as they had the best performance in the previous analysis.

To examine the UAR, we note that each model has an optimal cutoff for the predictions. We set the optimal cutoff at 0.45, slightly lower than 0.5, to quantify the slight unfairness of the field being studied. This means that a manager that has a 46% chance of being sacked will be considered sacked in the models.

Our results show that the most complex models fitted performs optimally in these circumstances with a UAR of 0.75, as shown by the UAR values for each model in Table 1. However, we have also observed that when we increase the optimal cutoff, the most naive model performs better than the more complex ones. This is expected, as if we are trying to base the sacking of a manager solely on results, the threshold is higher, as decision makers seem to not consider this the most important factor, as previously hypothesized, a sacking is more than losing a few games. Hence, the naïve models present missing features that we add progressively. Nonetheless, we have found that Complex Model #3, followed by Complex Model #2 are the most optimal models.

Table 3. UAR on unseen data.

Model	UAR of Test Set (0.45)	Number of features
Naïve model #2	0.72	1
Complex Model #1	0.70	28
Complex Model #2	0.73	59
Complex Model #3	0.75	67

2.3 Determinants of a Dismissal

We now focus on understanding the relevant determinants of sacking in football. We focus on five aspects: sentiment of the fans, match statistics, manager as a manager, manager as a player, and human bias.

Sentiment Analysis. The addition of sentiment into the model has dramatically reduced the number of important features from the previous model. Figure 1(a) shows that Negative Vader Sentiment on the Club and Negative TextBlob Sentiment on the Manager are both relevant determinants of a dismissal. As the negative sentiment increases, the probability of being sacked

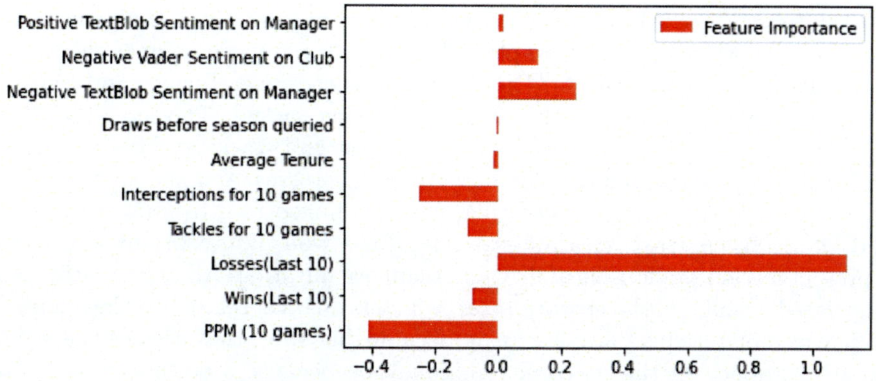

(a) Complex Model #3: Relative importance of features

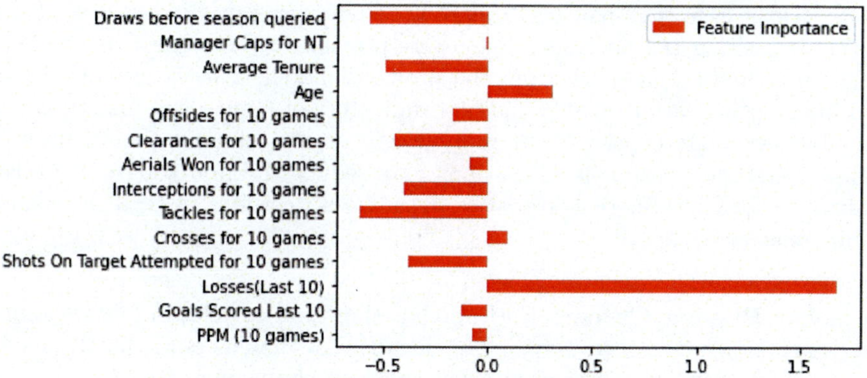

(b) Complex Model #2: Relative Importance of Features

Fig. 1. Relative importance of features for best models

increases as well. This is in line with the discussion in the introduction which fans' anger towards the club and manager have an influence on a dismissal. Furthermore, the sentiment towards the manager has more of a weight on the outcome compared to the sentiment towards the club, which was equally assumed.

We cannot definitely conclude which of the two sentiment packages, TextBlob and Vader, is definitely better for our purpose. The results would suggest that they perform differently on tweets addressing the manager and the club, as the Negative Sentiment on the Club with TextBlob and the Negative Sentiment on the manager with Vader are both rendered irrelevant. It would be viable to explore the type of language used overwhelmingly in the respective categories. We also observe that Positive Sentiment is rendered almost irrelevant. We can hypothesize this is due to high displease (negative sentiment) drowning out the low praise (positive sentiment), which makes the sentiment closer to neutral.

Match Statistics. Many of the match statistics had been rendered irrelevant in the Complex Model #3. In Complex Model #2 however, the higher the average of Offsides, Clearances, Aerial Duals won, Interceptions, Tackles and Shots on Goal over the 10 games, the less chance of being sacked. The most significant determinants are the Clearances, Interceptions and Tackles. This is coherent with real world assumptions. The more offsides, crosses, and shots a team has a game, the more chances it is creating, which is linked to attacking performance and effort. In contrast, clearances, aerial duels won, interceptions and rackles represent a team's defensive ability and team organisation. All these statistics are associated with high player work rates, which is directly related to the manager's ability to motivate his players. It is noticeable, however, that crosses has a slight negative impact on the outcome. This may be because a cross can be a 'lazy' form of attack, instead of intricate passing.

In Complex Model #3 with the sentiment scores, the average tackles and interceptions over the span of 10 games stay determinants in the model, negatively impacting the outcome. They were previously the highest determinants in the match stats category from the previous model. This is because Tackles and interceptions are probably the best indicators of team effort and High work rate. However, interceptions have more weight on the outcome then tackles. We can say that the sentiment of the fans may reduce the prerogative of decision makers to look at different statistics to make a decision, instead focusing on public relations.

Manager Playing Career. Both models shows no correlation with the manager's previous career as a player. This means that managers are mostly held to the same standard as others even with a glamorous playing career.

Manager Managerial Career. According to our models, the manager's previous career is rendered irrelevant with sentiment but without, the average tenure and draws in his career have a negative weight on the outcome of being sacked. A longer average tenure is representative of a non-problematic and successful manager, whereas the number of draws the manager has culminated in his career is synonymous with not losing, and it is equally important in a relegation or title fight, as a draw is much better than a loss. In both models however, the most relevant metrics of a manager's career are the ones from the previous 10 games, and they are the most important determinants unsurprisingly. The number of losses suffered in the last 10 games is the most important determinant in our model, it has four times the weight of any other feature's importance in both models. The more games the manager has lost in the last 10, the higher the chance he has of being sacked. As opposed to the other determinants, the significant weight shows that this chance of being sacked is far superior to the others. The PPM for 10 games, is the driving determinant in our well performing naive model, Naive model #2, but loses its influence when we consider the losses, but regains some influence when including the sentiment in the model. This is in line with the naive presupposition that the more points a manager gains over

10 games, the less chance he has of being sacked. It proves to not be as naive as expected.

Human Bias. Interestingly, older managers seem to have an increased chance of being sacked. It was assumed that older age increased respect that decision makers have towards the manager. However, it seems that being older is less of a benefit than being younger. One possible hypothesis is that a younger manager brings more energy and can be instrumental towards a long term project, whereas an older manager can be perceived as a short term plan. It is rendered irrelevant however by the sentiment as the opinion of the fans renders this bias obsolete.

3 Conclusions

This paper investigated the determinants of managerial turnover in football. We collected statistics from England's top two divisions and tweets over four seasons. In addition to traditional performance indicators, we examined whether *social pressure* contributes to dismissals, or could be used to predict them. We used sentiment analysis on tweets as a proxy for the public opinion towards clubs and managers. We successfully created elasticnet regularised logistic regression models obtaining precision, recall, accuracy, and F1 scores over 0.80. The unweighted average recall (UAR) scores are over 0.70, with a maximum of 0.75, suggesting a good generalisation. As expected, managers' *points per match* and *losses* over the last ten games are the driving factors of a dismissal. Nonetheless, the model suggests that the fans anger towards the club and the manager could be the tipping point in a dismissal decision, and that team effort also weights into the decision (e.g., the number of *interceptions* and *tackles*). Hence, even if managers will mostly be sacked because of the number of losses and their lack of point in recent matches, the significance of social pressure indicators suggests a complex decision scenario. It is possible that a negative sentiment towards managers is an emergent collective effect driven by bad performance; It could also be driven by prejudice or previous encounters as an opponent. This study does not address the causes of this social pressure and further investigation is required.

References

1. Allen, W.D., Chadwick, C.: Performance, expectations, and managerial dismissal: evidence from the national football league. J. Sports Econ. **13**(4), 337–363 (2012)
2. Audas, R., Dobson, S., Goddard, J.: Organizational performance and managerial turnover. Manag. Decis. Econ. **20**(6), 305–318 (1999)
3. Bachan, R., Reilly, B., Witt, R.: The hazard of being an English football league manager: empirical estimates for three recent league seasons. J. Oper. Res. Soc. **59**(7), 884–891 (2008)
4. Barros, C.P., Frick, B., Passos, J.: Coaching for survival: the hazards of head coach careers in the German 'bundesliga'. Appl. Econ. **41**(25), 3303–3311 (2009)

5. Calvin, M.: Living on the Volcano: The Secrets of Surviving as a Football Manager. Random House (2015)
6. Chase, H.W., Glickman, M.E.: The analytics of getting sacked: coach firings in the national football league. Significance **13**(4), 34–37 (2016)
7. de Dios Tena, J., Forrest, D.: Within-season dismissal of football coaches: statistical analysis of causes and consequences. Eur. J. Oper. Res. **181**(1), 362–373 (2007)
8. d'Addona, S., Kind, A.: Forced manager turnovers in English soccer leagues: a long-term perspective. J. Sports Econ. **15**(2), 150–179 (2014)
9. Fizel, J.L., D'Itri, M.P.: Managerial efficiency, managerial succession and organizational performance. Manag. Decis. Econ. **18**(4), 295–308 (1997)
10. Frick, B., Barros, C.P., Prinz, J.: Analysing head coach dismissals in the German "bundesliga" with a mixed logit approach. Eur. J. Oper. Res. **200**(1), 151–159 (2010)
11. Godin, F., Zuallaert, J., Vandersmissen, B., De Neve, W., Van de Walle, R.: Beating the bookmakers: leveraging statistics and twitter microposts for predicting soccer results. In: KDD Workshop on Large-Scale Sports Analytics, New York, NY, USA, pp. 2–14. ACM (2014)
12. Kumar, A., Roshan, D.: Football market size, share & growth: Industry forecast - 2027, May 2021. https://www.alliedmarketresearch.com/football-market-A11328
13. Lange, D.: Football clubs revenue ranking 2019/20, October 2021. https://www.statista.com/statistics/271581/revenue-of-soccer-clubs-worldwide/
14. Pacheco, D., de Lima Neto, F.B., Moyano, L., Menezes, R.: Football conversations: what twitter reveals about the 2014 world cup. In: Anais do IV Brazilian Workshop on Social Network Analysis and Mining. SBC (2015)
15. Pacheco, D.F., Pinheiro, D., de Lima Neto, F.B., Ribeiro, E., Menezes, R.: Characterization of football supporters from twitter conversations. In: WI, pp. 169–176 (2016)
16. Price, J., Farrington, N., Hall, L.: Changing the game? the impact of twitter on relationships between football clubs, supporters and the sports media. Soccer Soc. **14**(4), 446–461 (2013)
17. Rieple, A., Vyakarnam, S.: The case for managerial ruthlessness 1. Br. J. Manag. **7**(1), 17–33 (1996)
18. Rowe, W.G., Cannella, A.A., Jr., Rankin, D., Gorman, D.: Leader succession and organizational performance: Integrating the common-sense, ritual scapegoating, and vicious-circle succession theories. Leadersh. Q. **16**(2), 197–219 (2005)
19. Schumaker, R.P., Jarmoszko, A.T., Labedz, C.S., Jr.: Predicting wins and spread in the premier league using a sentiment analysis of twitter. Decis. Support Syst. **88**, 76–84 (2016)
20. New statistics reveal average tenure of managers in England. https://www.skysports.com/football/news/11688/9875915/average-tenure-of-managers-in-england-just-123-years
21. Smith, A.: Football manager job security at all-time low, May 2020. https://www.skysports.com/football/news/11096/10811856/football-manager-job-security-at-all-time-low-sky-sports-study-finds
22. Thornthwaite, L.: Social media and dismissal: towards a reasonable expectation of privacy? J. Ind. Relat. **60**(1), 119–136 (2018)
23. Van Ours, J.C., Van Tuijl, M.A.: In-season head-coach dismissals and the performance of professional football teams. Econ. Inq. **54**(1), 591–604 (2016)

Nature vs. Nurture in Science: The Effect of Researchers Segregation on Papers' Citation Histories

Ana Maria Jaramillo[1(✉)], Felipe Montes[2], and Ronaldo Menezes[1,3]

[1] BioComplex Laboratory, Computer Science, University of Exeter, Exeter, UK
ajaramillo@biocomplex.org, r.menezes@exeter.ac.uk

[2] Department of Industrial Engineering, Universidad de los Andes, Bogotá, Colombia
fel-mont@uniandes.edu.co

[3] Computer Science, Federal University of Ceará, Fortaleza, Brazil

Abstract. Academia is a competitive world where researchers are judged by their productivity, and their strategies to get visibility and success (i.e., number of citations) vary. In addition to conducting rigorous research, there are social strategies that influence authors' and their papers' level of citations. The author's position in the co-authorship network affects the success of their papers. Hence, we want to understand if the authors' segregation in the co-authorship network relates to citations gained by a paper over time. We address this question by examining the patterns in Computer Science from 1975 to 2015 (and citations until 2020) from the Semantic Scholar Open Research Corpus. Specifically, we identify communities in the co-authorship network and classify them into segregation categories and core positions. Then, we compare the citation histories of papers written in those communities. We examine papers written solely by members of the same community (internal) and different communities (external), resulting in the following five categories: internal highly-segregated, internal non-segregated, external highly-segregated, external non-segregated, and external mixed. Our results show that from 1998 to 2010, internal highly-segregated papers gained fewer citations than internal non-segregated and external mixed papers. Also, from 2010 to 2015, external mixed papers gained more citations than internal non-segregated papers and even more citations than internal highly-segregated papers. We also found that in the network nucleus (from core decomposition), there is little difference in the citations of internal non- and highly-segregated papers. In contrast, in the network's periphery, internal non-segregated papers tend to gain more citations than internal highly-segregated papers since 2005. From this work, we conclude that papers written by a more diverse set of authors (measured by their network connectivity) receive more citations over time and that to compensate for the lack of diversity, their authors should be in central positions of the co-authorship network. Hence, this work could incentivise diverse co-authorships and strengthen researchers' cohesion to increase their papers' success.

Keywords: Science of science · Co-authorship networks · Citations attainment

A. S. Teixeira et al. (Eds.): CompleNet 2023, SPCOM, pp. 141–154, 2023.
https://doi.org/10.1007/978-3-031-28276-8_13

1 Introduction

The amount of scientific works written in the last couple of years exceeds scientists' reading capacity, which has increased the challenges for young researchers to get successful careers and well-cited papers [28]. It is more difficult for researchers without well-established careers to explore new ideas and create innovative/disruptive research because of the fast pace at which literature increases [29, 39]. Much work has been done to analyse the behaviour of citations with scientometrics and bibliometric analyses, and some conclusions have emerged related to the social biases affecting the citation processes. There is strong evidence about how researchers with specific demographics such as gender [13,37] or countries of affiliation [12,36] get more cited. Also, there is work regarding strategies in the social processes involved in research activities, such as the first mover advantage [16,24], the selection of co-authors with implications in their position in the co-authorship networks [25], and the relationship with citations gained from inside or outside their disciplines [19].

Co-authorship networks are formed by nodes as authors and links connecting authors when they co-author a paper. In some disciplines, these networks are prone to increase their clustering coefficient [21,27] because the scientific practices have moved from works written by single researchers to teams [39], which form communities of highly collaborative researchers [22]. Co-authorship networks have a clear relationship with citation rates. Particularly in STEM disciplines such as Computer Science and Physics, there is evidence of the high rates of self-citations [40], the increment of citations done by previous co-authors [42], more citations when researchers are located in central positions of the co-authorship network [5,35]. Also, in our previous work in Computer Science, we found that researchers in highly-segregated communities receive more citations from members of their same community [14].

The number of citations as a measure of success can be criticised because not all citations are supportive (can discredit or correct the work) [1,32], and a substantial number of citations are done without direct relation to the current work [31,38]. Nevertheless, the number of citations can be seen as a proxy of the scientific community's collective wisdom about a piece of work [6]. Also, in current recommendation systems (e.g., Elsevier, Google Scholar), the algorithms favour papers with more citations [4], reinforcing their visibility and increasing algorithm biases [26]. Lower visibility of minorities [9] and their under-representation in top-ranked positions [15], as well as increments in misconceptions and stereotypes over the most vulnerable [26], have been related to those recommendation algorithms. In addition, using the number of citations to train the algorithms reduces access to diverse information, knowledge, and epistemologies [18], closing the loop and affecting co-authorship networks. Hence, researchers affiliated with institutions in the global south receive fewer citations and are located towards peripheral positions in scientific networks [11,29]. In social media, for example, recommendation algorithms have effects on the networks' growth, clustering coefficient, homophily, and community structure [10,34]. However, how the content produced in those communities (e.g. published papers) is affecting their metrics of visibility (e.g. the number of citations) over time still needs to be studied.

The analyses here extend our previous work where highly-segregated communities were defined in co-authorship networks of Computer Science, and their impact on overall citations received by their members was studied [14]. In that study, we use Girvan & Newman's definition of community as a group of vertices in the network with dense connections within the group and only sparser connections with different groups [23]. The results showed that during 2006, 2010 and 2014, and when compared to researchers in non-segregated communities, researchers in highly-segregated communities received more citations when located in the nucleus or periphery (not in middle cores) and also received a higher proportion of those citations from members of their same community.

Here, we study papers of Computer Science written from 1975 until 2015, with citation histories analysed until 2020. Citation histories is a term quoted by Parolo et al. [30], referring to the citations gained by a paper in a time interval. In the current paper, citation histories refer to the cumulative total citations that a paper receives over time. We want to understand how the citation histories of those papers, and consequently their visibility, are affected by the segregation category of the community's authors at the time of publication. Then, our analysis has three components: *(i)* We classify communities of co-authors on segregation and core categories. *(ii)* We classify papers as internal (written exclusively by authors of the same community) or external (written by authors of different communities). *(iii)* We analyse citation histories of internal and external papers written by authors of non- and highly-segregated communities in the nucleus and periphery of the co-authorship network.

Our results show that from 1998 to 2010, internal highly-segregated papers (those written within highly-segregated communities) ended their citation histories with an average of 10 fewer citations than *(i)* internal non-segregated papers (those written within non-segregated communities), and *(ii)* external mixed papers (those written by co-authors in communities of different segregation categories). Also, from 2010 to 2015, external mixed papers gained more citations than internal non- and highly-segregated papers. When we separate the communities by their core position, we found no difference in the citations of internal non- and highly-segregated papers in the nucleus. In contrast, in the network's periphery, internal non-segregated papers have more citations than internal highly-segregated papers since 2005.

The paper is organised as follows: Sect. 2 describes the dataset, co-authorship network, community partition, and definition of segregation and core categories. Section 3 shows the definition of internal and external papers and their classification into internal non- and highly-segregated papers and external non-, highly-, and mixed papers. In Sect. 4, we show the citation histories for all the papers in our dataset, and we compare the citation histories of papers from the previous categories. Finally, Sect. 5 discusses our main contributions, limitations and final remarks.

2 Data and Methods

We analyse the relationship of papers forming communities in different segregation levels and core positions with the papers' citation histories for Computer Science. We consider the Semantic Scholar Open Research Corpus [17] from 1975 to 2015 to have a frame in which the discipline was consolidated and have at least 5 years for receiving citations. Our data goes until 2020, but we do not analyse the last 5 years as more time needs to be passed to analyse those papers' citation histories. We build a co-authorship network for each year from the papers available in the dataset and analyse their community structures; however, for the citation histories analysis, we display the results of 3 years in particular (2005, 2010, 2015). The choice is somewhat arbitrary but also done due to space limitations; our goal is to show the general trends. Still, in interpreting our results, we write about the complete longitudinal analysis across all years.

For this analysis, we study each network's Largest Connected Component (LCC), and Table 1 shows the co-authorship network LCC characteristics for the three years. A node represents a researcher, and a link forms when two researchers co-author at least one scientific publication in the year. The links are weighted using the strength of the pairwise co-authorship, defined as the sum of common publications, dividing each publication by the number of co-authors [7, 25]. We found the community partition for each co-authorship network using the Label-propagation algorithm [33]. We group the communities into different categories of segregation and core position based on the methodology of previous work [14].

Table 1. Characteristics of the co-authorship network Largest Connected Component in 2005, 2010, and 2015. The communities were detected with the Label-propagation algorithm [33].

Metric per year	2005	2010	2015
Number of papers	99,956	184,642	303,550
Number of communities ($\geq$ 3 researchers)	20,896	39,998	57,041
Number of researchers in communities ($\geq$ 3 researchers)	211,591	407,532	589,574
Number of internal community papers	74,078	128,415	190,327

First, we compute segregation using the Spectral Segregation Index (SSI) [8], and we study its probability density function (PDF), mean (μ) and standard deviation (σ). We group the communities into three categories: non- (SSI $\leq \mu - \sigma$), medium- ($\mu - \sigma <$ SSI $< \mu + \sigma$), and highly-segregated ($\mu + \sigma \leq$ SSI). In Fig. 2**A**, we show the PDFs for all the years and in Fig. 4 left panel, we show an example of non- and highly-segregated communities. We can see how the distributions have their largest peaks towards high values of SSI. The distribution of SSI gets less skewed when the year increases, and fewer communities are at the peak of the curve. We infer this small flattening is because the co-authorship networks get more interconnected over time, and the communities segregate less.

Second, we compute the relative core position of the communities because previous work showed the relationship of the segregation category with the core position of each community [14]. For each year, we create a network in which each community is a node, and links between these nodes exist if the members share co-authorships. Then, we apply the k-core decomposition algorithm [3] and assign each community to a correspondent core. As shown in Fig. 1, the k value corresponds to the number of cores, and it differs for each year. The figure shows how a trend emerges over the years: highly-segregated communities in red are located towards the periphery, while non-segregated communities in blue are located towards the nucleus. Then, we compute the relative value of each core to compare different years normalising from 0 (periphery) to 1 (nucleus with the maximum core). In Fig. 2**B**, we show the PDFs of the relative core for all the years. We can see how most of the communities are in the network's periphery, with a smaller peak in the nucleus of the communities network. In this case, there is not a clear trend of the PDFs to decrease their peak values, which we infer is because of the preferential attachment tendency in this particular network: researchers with high connectivity in the nucleus will attain more connectivity and researchers in the periphery can be completely new and not expose to those researchers in central cores. Here, we separate the communities into three categories for the current analysis: nucleus (maximum k-core value), periphery (minimum k-core value), and middle cores (cores that are in between the nucleus and the periphery).

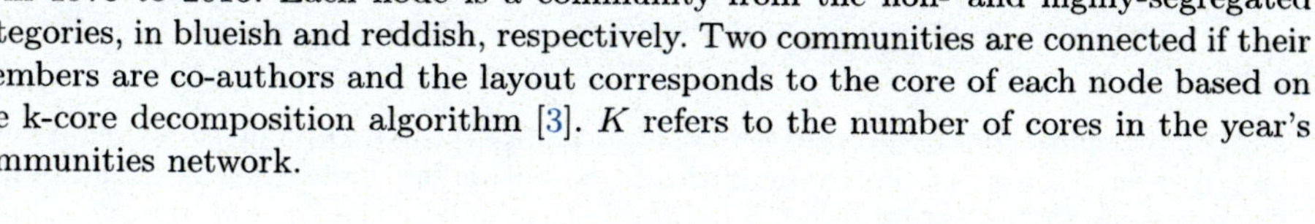

Fig. 1. Networks of communities with k-shell visualisation with five years of difference from 1975 to 2015. Each node is a community from the non- and highly-segregated categories, in blueish and reddish, respectively. Two communities are connected if their members are co-authors and the layout corresponds to the core of each node based on the k-core decomposition algorithm [3]. K refers to the number of cores in the year's communities network.

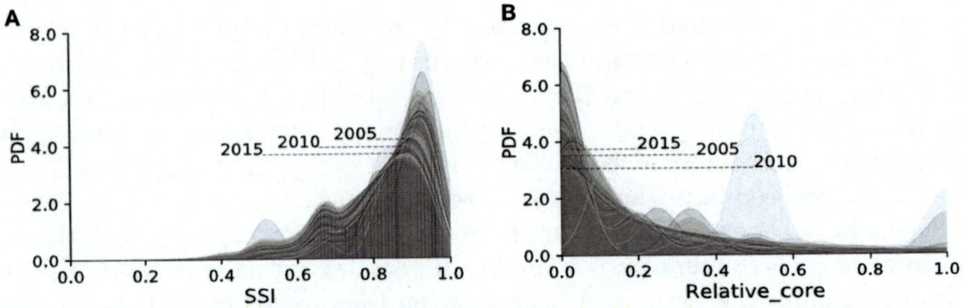

Fig. 2. Probability density functions (PDFs) for **A** the Segregation Spectral Index (SSI) of the communities and **B** the relative core of the communities in the network of communities, both metrics are calculated using the co-authorship networks from 1975 to 2020. We highlight the curves of the studied years in this manuscript to guide the reader.

Summarising, we group the communities into three categories of segregation and three categories of core position. In Fig. 3, we show plots for the proportion of communities in each category from 1975 to 2015. The SSI distributions have a slight flattening (Fig. 2**A**) over the years. The proportion of communities in each category of segregation has low variability (Fig. 3**A**): most of the communities are in the medium-segregated category, and non- and highly-segregated communities have a similar proportion of communities. On the contrary, the relative core distribution has similar curves over the years (Fig. 2**B**), but the proportion of communities in the nucleus remains low while there is high variability in the proportion of communities in the periphery (Fig. 3 **B**).

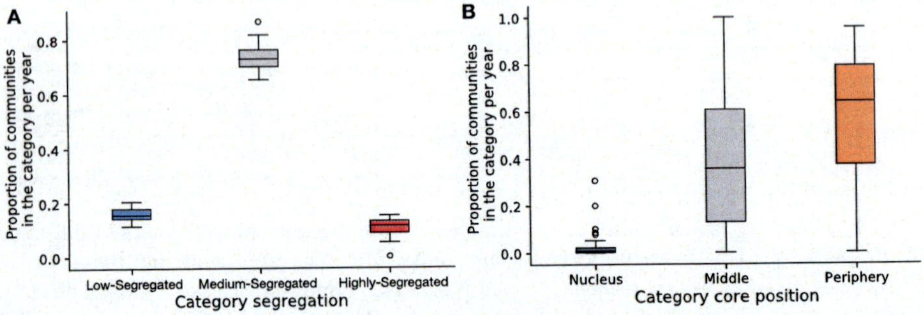

Fig. 3. Box plots for the proportion of communities in each category of **A** Segregation and **B** core position of the communities in co-authorship networks LCC of Compute Science from 1975 to 2015.

3 Categories of Papers

We analyse the citation histories of papers in our dataset and their relationship with being written in communities of different categories of segregation and core positions. In this paper, we focus exclusively on non- and highly-segregated communities because we want to understand the limits of the SSI spectrum, and we let the analysis of medium-segregated communities for future work. Hence, we define the five categories of papers listed in Fig. 4 middle panel. Henceforth, we refer to "internal" papers as those written by members of the same community. And we refer to "external" papers as those written by authors from different communities. For external papers, we have three options: authors in different communities that are non-segregated, highly-segregated, or mixed (non- and highly-segregated). In the case of the external papers, we do not differentiate by the percentage of authors in each community. For example, if a paper with 7 authors has 6 authors in highly-segregated communities and 1 author in a non-segregated community, we consider this paper "external mixed". The right panel of Fig. 4 shows the distribution of the proportion of papers in each one of the five categories per year. We can see how a larger proportion of the papers is written

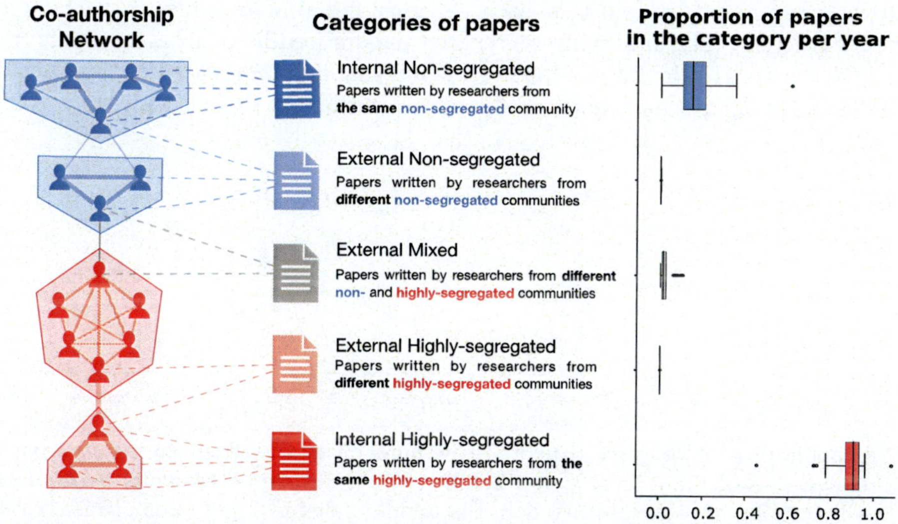

Fig. 4. Definition of the categories of papers used in this analysis. The left panel shows a sample from the co-authorship network with nodes grouped based on the community partition. From top to bottom, the first two groups in blue refer to non-segregated communities, and the second two groups in red refer to highly-segregated communities. Links in colour refer to those co-authorships inside the community, while those links in grey refer to co-authorships across different communities. The middle panel refers to the 5 categories of papers that we analyse. Dashed links between both panels refer to the authors (left) of a paper (middle). The right panel shows box plots for the distribution of the proportion of papers in each category per year in the same order as the categories of the middle panel.

inside highly-segregated communities (internal highly-segregated), followed by papers written by authors in different segregated communities (external highly-segregated).

4 Citation Histories of Papers

We want to understand the relationship between papers written in communities with different segregation categories and core positions in the network and their gained citations over the years. Then, we study the citation histories of the papers, referred to as the cumulative total citations that a paper receives from being written in the year y until 2020. For each year, we compute the mean μ and the interval confidence of $\pm\ \alpha = 0.05$ for the cumulative citations gained by papers in each category. As we see in Fig. 5, internal highly-segregated papers gain fewer citations than internal non-segregated and external mixed papers over time in the three years of analysis, a trend that repeats all years since 2005. We also can see in **B** and **C** that external mixed papers gain more citations than internal non-segregated papers, a trend that has repeated all years since 2010. Also, we find that the highest increment of the citation histories is around 4 years, then we highlight the number of cumulative citations in the 4th year for all the panels, and we find that each category has similar values. For example, in the 4th year, internal highly-segregated papers (reddish curves in the figure) have 5 or fewer citations since 1994, while internal non-segregated papers (bluish curves in the figure) have between 5 and 10 citations in the 4th year since 1990.

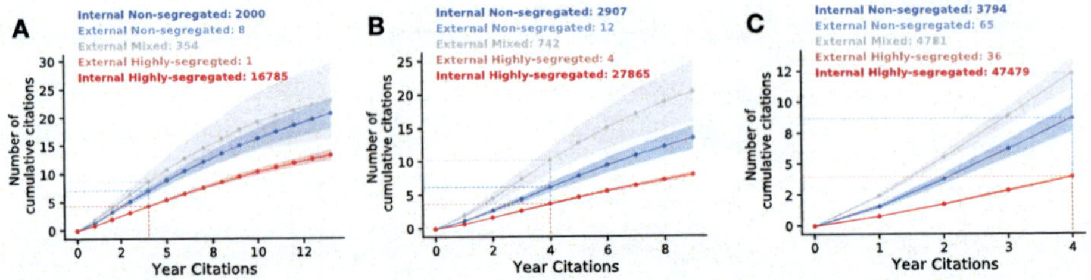

Fig. 5. Citation histories for papers of Computer Science written during **A** 2005, **B** 2010, and **C** 2015 from 5 categories based on their authors' community segregation. The description of each category is in the middle panel in Fig. 4. Complementary to Fig. 4 left panel, the number of external papers from authors in communities of the same segregation category is very small to analyse their citation histories and compare them with the other categories. Then, we do not plot the citation histories of external non- or highly-segregated papers. Here, the y-axis represents the number of cumulative citations per paper, the x-axis represents the number of years since publication, and the curves show the average value of the cumulative citations.

We want to study the citation histories of papers written in communities with different segregation categories and core positions. Then, we compare the citation histories of the internal papers written in non- and highly-segregated communities in the nucleus and periphery of the communities network. As we see in Fig. 6, there are not enough papers in the nucleus to have consistent results, but apparently, there is no difference in the citation trends for communities in different segregation categories (**A-C**). On the contrary, in the periphery, internal highly-segregated papers gain fewer citations than internal non-segregated papers (**D-F**), a trend that repeats all years since 2000. In addition, when we highlight the number of cumulative citations in the 4th year for all the panels, in the periphery (**D–F**), the trends are similar for internal highly-segregated papers having 5 or fewer citations and internal non-segregated papers having between 5 and 10 citations.

Considering the number of internal papers in non-segregated communities, their value grows slowly. However, the increment of papers in highly-segregated communities is faster (Fig. 4right). When separating by core position (Fig. 6), there are more papers in the periphery (bottom row) than in the nucleus (top row) for each segregation category. Looking at the figure legends, we see around 10 times more internal highly-segregated than internal non-segregated papers in the nucleus (820/84 in **C**). In the periphery, the number reaches about 15 times more (15653/987 in **F**).

In summary, papers that have gained the most cumulative citations since 2005 are those written by authors that are part of communities with different segregation categories, followed by papers written in non-segregated communities. In addition, the difference between papers written in non- and highly-segregated communities occurs in the network's periphery, but there is no such difference in the nucleus.

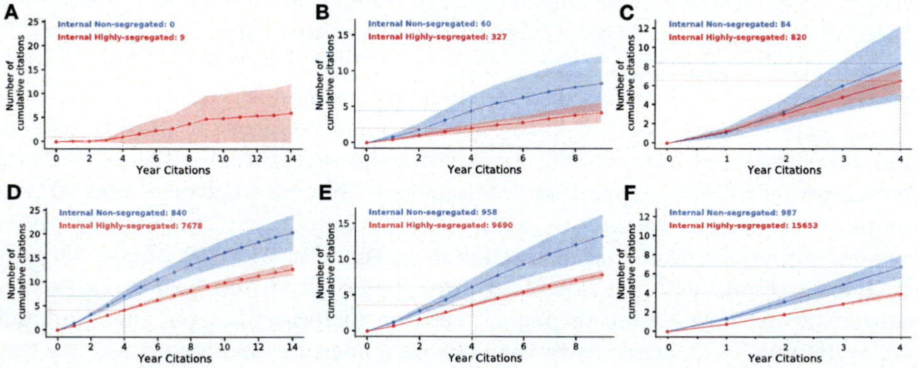

Fig. 6. Citation histories for papers of Computer Science written during **A** and **D** 2005, **B** and **E** 2010, and **C** and **F** 2015, and being internal to the two categories of segregation: non- and highly-segregated communities. The top row refers to papers written in communities located in the network's nucleus, and the bottom row refers to papers written in communities in the periphery. Here, the y-axis represents the number of cumulative citations per paper, the x-axis represents the number of years since publication, and the curves show the average value of the cumulative citations.

5 Discussion

Citation counts have been used to evaluate researchers' performance. It has a high role in the visibility of researchers in academic search engines because ranking algorithms use this metric to prioritise the importance of academic work [4]. Then, understanding the behaviour of the citations that a paper receives over time is paramount to studying trends in scientific practices, such as collaboration patterns or topics of research [20]. Prior studies analysed how the co-authorship patterns shape the position of papers in citation networks and found a relationship between having a larger number of previous co-authors and having more citations per year [5,35]. In addition, in previous work, we found how researchers immersed in highly-segregated communities tend to receive more citations in the nucleus of the co-authorship network and more citations from their same community [14]. Here, we concentrate on studying the direct relationship between the segregation of authors with the citations that their papers receive over time (named "citation histories"). We compare the citation histories of papers written inside and across different non- and highly-segregated communities based on the communities' Spectral Segregation Index (SSI) distribution. Then, we analyse the citation histories for papers inside those communities by their position in different cores of the communities' network. As a case study, we used literature on Computer Science in the Semantic Scholar Open Research Corpus from 1975 to 2015.

We analyse the citation histories of papers written by authors inside (internal) non- and highly-segregated communities and papers written by authors across different communities (external). In general, we found that papers written by a more diverse set of authors (measured by their network connectivity) receive more citations over time and that to compensate for the lack of diversity, their authors should be in central positions of the co-authorship network.

Specifically, our results show that from 1998 to 2010, internal highly-segregated papers got fewer citations than internal non-segregated and external mixed papers. From 2010 to 2015, external mixed papers outperformed internal non-segregated papers. When analysing the segregation category with the core position of the communities, we found no difference between internal non- and highly-segregated papers in the nucleus. In contrast, internal non-segregated papers outperform highly-segregated ones in the network's periphery. We combine both results with previous literature that showed how well-connected co-authorship networks improve papers' citation histories [42]. On the one hand, we argue that for researchers in the network's nucleus, the segregation category is not a concern because they have higher connectivity, giving them access to larger audiences than those in the periphery. And on the other hand, researchers in the periphery need to increase their connectivity with researchers that help to decrease their segregation levels to improve the citations of their papers.

Our results need further analysis as the larger number of internal papers written in highly-segregated communities could contain papers with different levels of relevance for the readers. Therefore, we could understand which characteristics of those papers help them tackle low citations, such as the topic and demographics of the authors, as these properties have been related to citation's gain [12,13,36,37].

Studies on the attention economy have exposed a problem in the growth of science [28]. Over time, more researchers publish more papers, but the time for reading and conducting research remains the same or has shrunk in some disciplines. As shown in our previous work, researchers in the periphery in highly-segregated communities have to publish more to increase their total citations and have similar values of total citations to researchers in non-segregated communities [14]. On average, the former publish 5 times more papers than the latter but gain fewer citations per paper because they do not collaborate with the most central researchers, having papers with less visibility and gaining less for more work. Here, we consider more work in terms of a higher number of papers, and our results should be read cautiously because we do not evaluate the quality of the papers or the quality of the publication venue; perhaps their discrepancies in predicting the number of citations [41]. In addition, internal non- and highly-segregated papers written in the nucleus do not show differences in their citations, and this can be due to researchers in highly-segregated communities in the nucleus showed a higher proportion of citations from their same community members in our previous work [14], which can be the reason of the even results.

6 Limitations and Future Work

First, our analysis does not differentiate the source of the citations to understand how diverse or interdisciplinary the audience of researchers citing the papers is. Further analysis should apply the techniques developed by bibliometricians to analyse the source of citations to correct rankings of researchers [1], avoid biases caused by self-citations [40], and understand how information flows across disciplines [19]. Second, we concentrate on one discipline: Computer Science; however, we do not differentiate by subfields which can have different citation and publication trends, nor correct by the impact factor of the publication venue. Further analyses are needed to understand which paper characteristics, e.g. author demographics, lead to a different impact under the same conditions of publication [13,37]. Finally, our third highlighted limitation is that we do not analyse how the transition of researchers from highly- to non-segregated communities impacts the citations that the papers gain over time and the corresponding effect of having time windows of one year. Researchers are mobile agents, and their co-authorship patterns change over time. Further analyses can be done to understand how academic mobility that increased citations from new collaborators [2] could affect papers published in non- and highly-segregated communities.

References

1. Aljuaid, H., Iftikhar, R., Ahmad, S., Asif, M., Tanvir Afzal, M.: Important citation identification using sentiment analysis of in-text citations. Telemat. Inf. (2021). https://doi.org/10.1016/j.tele.2020.101492
2. Aman, V.: A new bibliometric approach to measure knowledge transfer of internationally mobile scientists. Scientometrics **117**(1), 227–247 (2018). https://doi.org/10.1007/s11192-018-2864-x
3. Batagelj, V., Zaversnik, M.: An O(m) algorithm for cores decomposition of networks. arXiv:0310049 (2003)
4. Beel, J., Gipp, B.: Google scholar's ranking algorithm: an introductory overview. In: 12th International Conference on Scientometrics and Informetrics, ISSI 2009 (2009)
5. Biscaro, C., Giupponi, C.: Co-authorship and bibliographic coupling network effects on citations. PLoS ONE (2014). https://doi.org/10.1371/journal.pone.0099502
6. Bornmann, L., Marx, W.: The wisdom of citing scientists. J. Assoc. Inf. Sci. Technol. (2014). https://doi.org/10.1002/asi.23100
7. Cann, T.J.B., Weaver, I.S., Williams, H.T.P.: Is it correct to project and detect? assessing performance of community detection on unipartite projections of bipartite networks. In: Aiello, L.M., Cherifi, C., Cherifi, H., Lambiotte, R., Lió, P., Rocha, L.M. (eds.) COMPLEX NETWORKS 2018. SCI, vol. 812, pp. 267–279. Springer, Cham (2019). https://doi.org/10.1007/978-3-030-05411-3_22
8. Echenique, F., Fryer, R.G.: A measure of segregation based on social interactions. Q. J. Econ. **122**, 441–485 (2007). https://doi.org/10.1162/qjec.122.2.441
9. Espín-Noboa, L., Wagner, C., Strohmaier, M., Karimi, F.: Inequality and inequity in network-based ranking and recommendation algorithms. Sci. Rep. (2022). https://doi.org/10.1038/s41598-022-05434-1
10. Ferrara, A., Espin-Noboa, L., Karimi, F., Wagner, C.: Link recommendations: their impact on network structure and minorities. In: 14th ACM Web Science Conference 2022, WebSci 2022, pp. 228–238. Association for Computing Machinery, New York (2022). https://doi.org/10.1145/3501247.3531583
11. Gomez, C.J., Herman, A.C., Parigi, P.: Leading countries in global science increasingly receive more citations than other countries doing similar research. Nat. Human Behav. **6**(7), 919–929 (2022). https://doi.org/10.1038/s41562-022-01351-5
12. Gonzalez-Brambila, C.N., Reyes-Gonzalez, L., Veloso, F., Perez-Angón, M.A.: The scientific impact of developing nations. PLOS ONE **11**(3), e0151328 (2016). https://doi.org/10.1371/journal.pone.0151328
13. Huang, J., Gates, A.J., Sinatra, R., Barabási, A.L.: Historical comparison of gender inequality in scientific careers across countries and disciplines. Proc. Natl. Acad. Sci. United States Am. **117**(9), 4609–4616 (2020). https://doi.org/10.1073/pnas.1914221117
14. Jaramillo, A.M., Williams, H.T., Perra, N., Menezes, R.: The community structure of collaboration networks in computer science and its impact on scientific production and consumption. arXiv e-prints pp. arXiv-2207 (2022)
15. Karimi, F., Oliveira, M., Strohmaier, M.: Minorities in networks and algorithms (2022). https://doi.org/10.48550/ARXIV.2206.07113, https://arxiv.org/abs/2206.07113

16. Kong, H., Martin-Gutierrez, S., Karimi, F.: Influence of the first-mover advantage on the gender disparities in physics citations. Commun. Phys. **5**(1), 243 (2022). https://doi.org/10.1038/s42005-022-00997-x
17. Lo, K., Wang, L.L., Neumann, M., Kinney, R., Weld, D.: S2ORC: the semantic scholar open research corpus. In: Proceedings of the 58th Annual Meeting of the Association for Computational Linguistics (2020). https://doi.org/10.18653/v1/2020.acl-main.447
18. Lowrie, I.: Algorithmic rationality: epistemology and efficiency in the data sciences. Big Data Soc. (2017). https://doi.org/10.1177/2053951717700925
19. Lynn, F.B.: Diffusing through disciplines: insiders, outsiders, and socially influenced citation behavior. Soc. Forces **93**(1), 355–382 (2014). https://doi.org/10.1093/sf/sou069
20. Mukherjee, S., Romero, D.M., Jones, B., Uzzi, B.: The nearly universal link between the age of past knowledge and tomorrow's breakthroughs in science and technology: the hotspot. Sci. Adv. **3**(4), e1601315 (2017). https://doi.org/10.1126/sciadv.1601315, https://www.science.org/doi/abs/10.1126/sciadv.1601315
21. Newman, M.E.J.: Coauthorship networks and patterns of scientific collaboration. Proc. Natl. Acad. Sci. **101**(suppl_1), 5200–5205 (2004). https://doi.org/10.1073/pnas.0307545100, https://www.pnas.org/doi/abs/10.1073/pnas.0307545100
22. Newman, M.E.J.: Finding community structure in networks using the eigenvectors of matrices. Phys. Rev. E Stat. Nonlinear Soft Matter Phys. **74**(3 Pt 2), 36104 (2006). https://doi.org/10.1103/PhysRevE.74.036104
23. Newman, M.E.J., Girvan, M.: Finding and evaluating community structure in networks. Phys. Rev. E **69**, 026113 (2004). https://doi.org/10.1103/PhysRevE.69.026113
24. Newman, M.E.: The first-mover advantage in scientific publication. EPL (2009). https://doi.org/10.1209/0295-5075/86/68001
25. Newman, M.E.: Who is the best connected scientist?a study of scientific coauthorship networks. In: Complex Networks, pp. 337–370 (2004). https://doi.org/10.1007/978-3-540-44485-5_16
26. Noble, S.U.: Algorithms of Oppression. New York University Press, New York (2019). https://doi.org/10.2307/j.ctt1pwt9w5
27. Ortega, J.L.: Influence of co-authorship networks in the research impact: ego network analyses from microsoft academic search. J. Inf. (2014). https://doi.org/10.1016/j.joi.2014.07.001
28. Pan, R.K., Petersen, A.M., Pammolli, F., Fortunato, S.: The memory of science: inflation, myopia, and the knowledge network. J. Inf. **12**(3), 656–678 (2018). https://doi.org/10.1016/j.joi.2018.06.005
29. Park, M., Leahey, E., Funk, R.J.: Papers and patents are becoming less disruptive over time. Nature **613**(7942), 138–144 (2023). https://doi.org/10.1038/s41586-022-05543-x
30. Parolo, P.D.B., Pan, R.K., Ghosh, R., Huberman, B.A., Kaski, K., Fortunato, S.: Attention decay in science. J. Inf. (2015). https://doi.org/10.1016/j.joi.2015.07.006
31. Pavlovic, V., et al.: How accurate are citations of frequently cited papers in biomedical literature? Clin. Sci. (2021). https://doi.org/10.1042/CS20201573
32. Radicchi, F.: In science "there is no bad publicity": papers criticized in comments have high scientific impact. Sci. Rep. **2**, 815 (2012). https://doi.org/10.1038/srep00815
33. Raghavan, U.N., Albert, R., Kumara, S.: Near linear time algorithm to detect community structures in large-scale networks. Phys. Rev. E **76**(3) (2007). https://doi.org/10.1103/physreve.76.036106

34. Santos, F.P., Lelkes, Y., Levin, S.A.: Link recommendation algorithms and dynamics of polarization in online social networks. Proc. Natl. Acad. Sci. **118**(50), e2102141118 (2021). https://doi.org/10.1073/pnas.2102141118, https://www.pnas.org/doi/abs/10.1073/pnas.2102141118
35. Sarigöl, E., Pfitzner, R., Scholtes, I., Garas, A., Schweitzer, F.: Predicting scientific success based on coauthorship networks. EPJ Data Sci. **3**(1), 1–16 (2014). https://doi.org/10.1140/epjds/s13688-014-0009-x
36. Smith, M.J., Weinberger, C., Bruna, E.M., Allesina, S.: The scientific impact of nations: journal placement and citation performance. PLoS ONE **9**(10), 1–6 (2014). https://doi.org/10.1371/journal.pone.0109195
37. Sugimoto, C.R., Lariviere, V., Ni, C., Gingras, Y., Cronin, B.: Global gender disparities in science. Nature **504**, 211–213 (2013)
38. Teplitskiy, M., Duede, E., Menietti, M., Lakhani, K.: Why (almost) everything we know about citations is wrong: evidence from authors. In: STI 2018 Conference Proceedings (2018)
39. Uzzi, B., Mukherjee, S., Stringer, M., Jones, B.: Atypical combinations and scientific impact. Science **342**(6157), 468–472 (2013). https://doi.org/10.1126/science.1240474, https://science.sciencemag.org/content/342/6157/468
40. Van Noorden, R., Singh Chawla, D.: Hundreds of extreme self-citing scientists revealed in new database (2019). https://doi.org/10.1038/d41586-019-02479-7
41. Wang, D., Song, C., Barabási, A.L.: quantifying long-term scientific impact. Science **342**(6154), 127–132 (2013). https://doi.org/10.1126/science.1237825, https://www.science.org/doi/abs/10.1126/science.1237825
42. Zingg, C., Nanumyan, V., Schweitzer, F.: Citations driven by social connections? a multi-layer representation of coauthorship networks. Quant. Sci. Stud. **1**(4), 1493–1509 (2020). https://doi.org/10.1162/qss_a_00092

Using Vector Fields in the Modelling of Movements as Flows
A Case Study with Cattle Trade Networks

Sima Farokhnejad[1(✉)], Marcos Oliveira[1], Eraldo Ribeiro[2], and Ronaldo Menezes[1,3(✉)]

[1] Computer Science, University of Exeter, Exeter, England, UK
{sf503,m.a.oliveira,r.menezes}@exeter.ac.uk
[2] Computer Science, Florida Institute of Technology, Melbourne, FL, USA
eribeiro@fit.edu
[3] Computer Science, Federal University of Ceará, Fortaleza, Brazil

Abstract. Livestock production is one of the world's most important economic activities, involving nearly every country either as a producer or consumer. Indeed, around 1.3 billion people worldwide depend on livestock production. As livestock numbers increase due to the growth in the world population, so does the need for modelling and understanding the patterns in the movement of livestock, which is crucial for understanding global epidemic patterns. Interestingly, the structure of livestock movement is quite similar to other movements, such as human mobility, if we model the phenomena as cases of origin-destination (O-D) flows. Here, we introduce a methodology to better understand the dynamics of mobility patterns by characterising them as these flows while accounting for spatial information. Our approach looks into flows of movements as something that can be derived from networks. We demonstrate the power of our approach on cattle trading by examining a dataset from the Brazilian state of Minas Gerais, the country's largest cattle production. Our proposal is general and fits to any case in which the network is build from an O-D matrix.

Keywords: Origin-destination networks · Flow maps · Vector fields · Cattle epidemic modelling · Cattle trade networks

1 Introduction

Animal production is a major component of economies worldwide. It is accepted that animal production support 1 of every 5 people in the world [28], which is a demand that continues to increase as the planet's human population grows. 71 million tones of beef products were produced in 2018 by the top-6 producers in the world, namely, USA, Brazil, China, India, Argentina, and Australia. In 2020, the projected number of bovines in Latin America and the Caribbean was 0.5 billion and in North America and Europe was 0.3 billion [27].

A. S. Teixeira et al. (Eds.): CompleNet 2023, SPCOM, pp. 155–167, 2023.
https://doi.org/10.1007/978-3-031-28276-8_14

These large markets may suffer catastrophic losses when affected by contagious diseases. In 2021, the population of cattle and calves in the United Kingdom was approximately 9.44 million heads. However, this number was supposed to be larger because a foot-and-mouth (FMD) epidemic in 2001 resulted in the slaughter of approximately 3 million animals. The epidemic costed the UK agriculture industry around £3.1 billion but also affected other major sectors of the economy, such as the tourism industry, which suffered losses ranging from £140 million to £500 million a week [24].

To improve their disease-monitoring capabilities, many countries have invested in tools for tracking animal production, creating a market estimated at US$ 24 billion. For example, Australia established a livestock traceability system based on electronic ear tags for cattle management. Thailand tracks the inter-provincial movements of livestock by an online system called *e-movement* [33]. And since 2001, Japan has implemented a food card system based on the agricultural biographical system, which tracks agricultural production and marketing [35].

While current monitoring tools can provide an overview of important aspects of the animal product market, they might miss signs of animal diseases that can be hard to quantify because the losses might come from hidden sources. For example, while visible sources include the death of animals, decreased access to food, and poor quality of animal products, hidden losses might occur from a change in animal population structure, increased labour costs, and environmental issues such as CO_2 emissions.

There are many Network-Science approaches to model epidemics in various regions [3,11,17,19,21] using a variety of epidemiological approaches [16,23,34]. While the power of network approaches to model disease spread is undeniable, they suffer from shortcomings. Very few approaches have used large datasets such as the ones available in Brazil or the USA [12] where the level of uncertainty about the trade is quite high [2]. For example, Brazil is the second-largest producer of beef in the world with an annual production of 10.3 million metric tons, behind only the USA with 12.5 million metric tons and ahead of the European Union with 7.8 million tons [1]. Yet, as mentioned above, Brazil is known to have a less-than-perfect tracking system leading to uncertainty in the trade network [2].

Network-based approaches often ignore relevant spatial information; when cattle herds are transported between locations, areas in-between are affected. This information is lost in networks because they only capture origin-destination pairs. If the locations between the origins and destinations are not part of any trade, they are ignored in the network representation. However, they might have higher spatial centrality because ground transportation is used in most cattle movements. Figure 1 shows an example in which locations A, B, C, and D are origins and destinations, but W and Z are not. In the network representation (on the left), W and Z are not even nodes, and a network analysis would overlook that the flow of cattle passes through those locations (e.g., the shortest-paths movements). In the representation on the right, we can see the betweenness of the locations in which W and Z have central positions in the spatial movements.

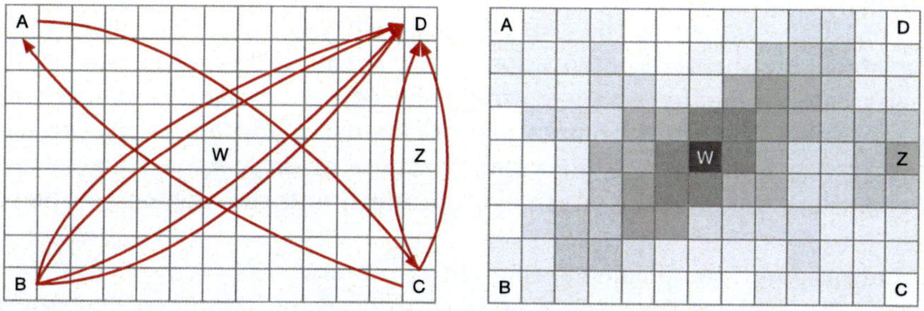

Fig. 1. Network modelling might miss important locations. In this toy example we have several movements represented as edges in a spatial network: (A→C), (C→A), 4 × (B→D), and 2 × (C→D). On the network representation on the left, the locations W and Z are not part of the representation because they are not involved in movements. However, the representation on the right has the potential to point out that locations W and Z are important as a confluence of spatial movements from several locations.

Hence, we propose to convert the network of movements into a vector-field representation, as it complements the benefits of network analyses. Furthermore, when datasets lack information (e.g., noise, uncertainty), our approach is more intuitive in adding locations based on vector fields than adding a network node or link. Last, our proposal is less computationally intensive than network algorithms, in particular, algorithms to analyse the dynamics of the network. Our approach is tested on a very large dataset from Brazil. We focus on cattle trading and examine a dataset from the Brazilian state of Minas Gerais, the country's largest cattle production. Our results show that our model provides an adequate representation that allows us to see the dynamics of trade and the effect of uncertainty even when datasets are massive.

2 Related Works

The understanding of cattle trade and its modelling using network science has caught the attention of researchers, given that networks are an excellent framework for capturing the structure of connections and cattle trade is primarily an origin-destination phenomenon in which cattle herds are moved around for various purposes: sales, fattening, slaughtering, to name a few. Networks can capture the structure of this economic activity, which in turn can be used to predict possible epidemic behaviours such as FMD disease.

Several works appeared in the literature after the FMD outbreak of 2001 when issues related to managing and controlling of infectious diseases in livestock were raised [5,13,14]. These works employed a network modelling that traces the transmission of diseases using the contacts between livestock locations that arise from trade activities.

After 2005, the use of the techniques from network science became more commonplace, with many works making use of open datasets of livestock movement [3,7,20–22,25]. Generally, the papers agreed with the fact that movement and the contacts are the main reasons for disease spread in livestock, especially cattle. As cattle production keeps growing it becomes necessary to understand the characteristics of livestock contact networks under conditions in the production countries. Most of the work done thus far looks at datasets in countries where the conditions for production are close to ideal, and the tracing can be easily done.

Our proposal complements the use of networks because it can capture dynamic mechanisms that would be hard to do using networks. Our approach can also be done computationally cheaper; hence, it is appropriate for large-scale datasets. We believe this is the first time this approach is used even though the literature mentions of flow field being used in migration patterns; these fields allow us to see various features in the fields such as saddle points, attracting areas, attracting nodes, and others [15]. For instance, Boyandin et al. [8] introduced a tool called JFlowMap for the analysis of flow maps, in particular the visualisation of such flows, yet, their work comprises of flows as networks and not as vector fields which is what we describe here. Their approach has been used to look at flows of specific kinds of movement, such as bicycles [10], showing that flows can be used for mobility data. Yet, such works are essentially network modelling representing the flows. We argue that vector fields are generic and that the methodology used here could very well be used in other spatial mobility data, such as human mobility [4].

In this paper, we use vector fields to model cattle trade flows. Our goal is to apply vector fields to other mobility data, but, for this, we have to use interpolation to make sure the field has a complete representation of the flow. Many tools exist to visualise vector fields, and a good comparison is provided by Laidlaw et al. [18]. Vector fields have been extensively used in weather, as they also provide better visualisations of the dynamics of such systems [30]. One of the main advantages is that one can use approaches for the identification of unusual/critical structures in these vector fields.

Figure 2 shows a few examples of critical structures that can be given a semantic interpretation in the context of epidemics and cattle movement. For instance, the "repelling focus" and "repelling node" structures can be considered possible sources of an epidemic, while an "attracting node" could be seen as the sink of an epidemic. That means that sources are rarely the destination of an outbreak. Vector fields allow for easier identification and visualisation of these substructures and how they change over time; the dynamics of sinks and sources can drive different interventions in the case of an epidemic; the intervention can be designed to deal with seasonality of movements if the vector flow reveals such patterns.

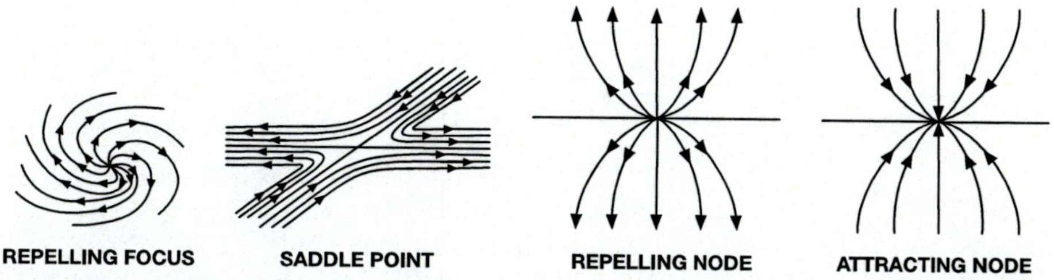

Fig. 2. Example of critical structures in vector fields. The use of vector fields allows for the identification of spatial critical points such as the ones in this figure. We can also look at the dynamics of these points overtime.

3 Methodology: Mobility Vector Fields

3.1 From Networks to Vector Fields

We indicated earlier that we want to go from an origin-destination representation (a network) to a vector field representation for the region being studied. Hence, the first step is to set divisions of the region being investigated—a granularity of the study—and project the network on this space. With this step, network nodes will be part of a particular cell in the spatial representation (Fig. 3A). With the projection in Fig. 3A, we can look into all the outgoing edges for each cell and taking them as vectors with distance and value, and then combine them into one vector joining pairs of cells. This is illustrated in Fig. 3A-1 and Fig. 3A-2 for one cell shown in grey. The set of vectors for the grey cell is combined in a way that only one vector exists between any pair of cells (yellow vectors in Fig. 3A-2). The second step is then to calculate the resulting vector from all the yellow vectors outgoing from a particular cell; this resulting vector shown in green in Fig. 3A-2 represents the general flow direction for that cell. Once this process is done for all cells with nodes with outgoing edges, we end up with what is shown in Fig. 3B. Note that only cells that have network nodes with outgoing edges have vectors, which means the vector field is incomplete. We will use an interpolation method to complete the field.

3.2 Vector Field Interpolation

As we have seen in the steps described in the previous section, the conversion of a network into a vector field, leads to a field where many locations in the studied area may not have a vector because they were not origins of for edges in the network (Fig. 3B). To assess the global behaviour of the vector field patterns, we need to complete the field with values that represent the flow in that location. In fact, in order to have an actual idea of flow, we have to interpolate in other points also. Interpolation methods involve constructing vectors for new points based on already-known vectors.

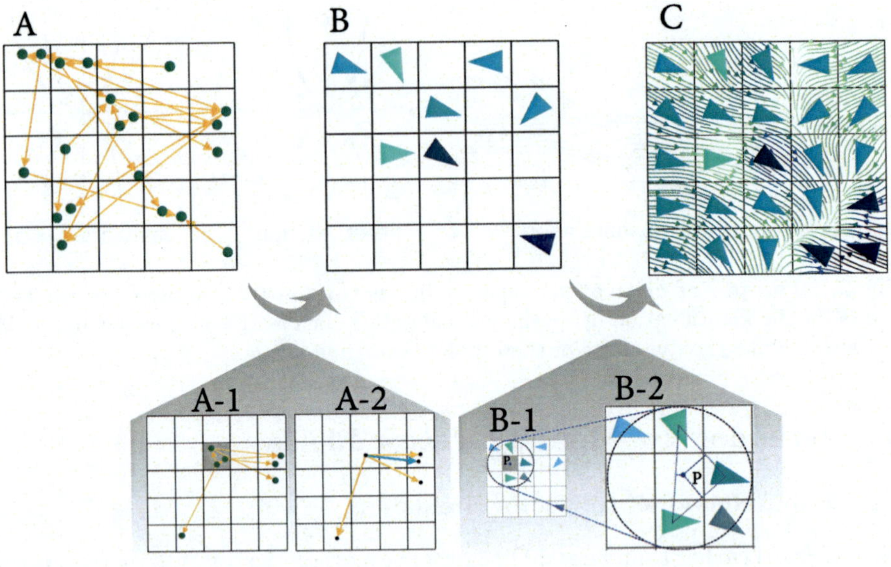

Fig. 3. The process of generating a vector field from a network. **(A)** a directed network representing movement is projected on a spatial map divided into parts. **(A-1)** focus of one cell/part of the space and the edges outgoing from that cell (in grey). **(A-2)** the movements are then converted into a resulting vector with the starting point at the centre of the origin cell (grey cell) and the end point at the centre of the destination cell (shown as the green vector). **(B)** all the resulting vectors from the network in (A) showing that many cells do not have a vector. **(B-1)** and **(B-2)** in order to estimate the vectors for each desired point in the grid, we implement a triangle-based interpolation method; the value of vector at a point P can be obtained by using the three nearest points with vectors. **(C)** interpolated vector field; different colours indicated different vector sizes.

There are multiple interpolation methods to employ when attempting to develop a continuous vector field. A triangle-based interpolation [31,32] is performed to estimate vector values at points in the grid where their vectors have not been determined. This method uses the known vectors of the three nearest points to calculate the vector for each point in space. As a result, to get the vector of point P in the grey cell in Fig. 3B-1, the three nearest points having vectors are used as illustrated in Fig. 3B-2. Assume that $\mathbf{v}_1$, $\mathbf{v}_2$, and $\mathbf{v}_3$ are the three nearest vectors. h_1, h_2, and h_3 represent the distance between point P and the side opposite the angle with the same indexed vector. Using triangle barycentric coordinates [29] and Delaunay triangulation on points with initial vectors, the vector at point P is determined using Eq. 1.

$$\mathbf{v}_P = \frac{h_1}{h_1 + h_2 + h_3}\mathbf{v}_1 + \frac{h_2}{h_1 + h_2 + h_3}\mathbf{v}_2 + \frac{h_3}{h_1 + h_2 + h_3}\mathbf{v}_3. \quad (1)$$

Once the triangulation is done for every point of interest, we end up with a complete vector field as depicted in Fig. 3C—this figure shows the main flows for each cell but also the flow for the entire field with more points being used.

3.3 Dynamics of Fields

We believe that the conversion from networks to vector fields leads to an interesting way of modelling the dynamics of mobility flows. In the case of cattle trade (Sect. 4), this is very useful because it allows us to observe seasonal patterns and locations that deserve more attention as they appear as features in the fields (Fig. 2) but also how these features change overtime.

The exploration of flow patterns globally, requires a concentration on a region (e.g., a part involved in an established division or a particular spatial part of a map) and observing how that region behaves overtime in terms of mobility. The dominant vector direction of a region, the way it changes overtime, and the similarity between the behaviour of vectors of different regions could be easily surveyed through a vector-field representation.

A cosine similarity approach is one possible method that allows us to determine the pattern of the flow direction originating from each region. The cosine similarity factor for each region could be calculated as

$$S_C(\mathbf{v}_k^i, \mathbf{v}_k^j) = \frac{\mathbf{v}_k^i \cdot \mathbf{v}_k^j}{\|\mathbf{v}_k^i\| \|\mathbf{v}_k^j\|}, \tag{2}$$

representing the dynamic auto-correlation of the vector for the region (spatial part) k, $\mathbf{v}_k$, in the time intervals i and j; this approach is adapted from the concept of dynamic correlation between two regions [6].

The dynamic analysis in the dominant flow directions can provide insightful information for decision makers. For this, flow directions are calculated over a set of pre-determined time periods. For instance, if we choose to use ten time intervals, we can generate ten vector fields using the process depicted in Fig. 3; the result is then the set of fields as seen in Fig. 4A-1.

Afterwards, the cosine value between each pair of consecutive time intervals is calculated using Eq. 2 to examine the changes in the vector related to each area of interest (i.e., a grid cell). In essence, this means that we have a feature vector for each area of interest, allowing us to cluster the areas based on feature-vector similarity. A k-means clustering method is used here; parts in the same group behave more similarly overtime than parts in other groups (Fig. 4A-2).

Clustering parts based on their mobility direction overtime helps us determine whether a part has an essential role in the global behaviour of mobility flow.

3.4 Identification of Critical Points

The analysis of vector fields can be accomplished by estimating it near some particular location. Vector fields have so-called critical points that reveal insights into their characteristics when we observe the global behaviour of the field near them. Critical points are points where the flow vanishes. Surrounding these points, the vector field has a distinct structure (Fig. 2).

Classification of critical points as sink, source, and saddle points is done according to the sign of the eigenvalues of the Jacobian matrix [26]; which is the matrix of all first-order partial derivatives of the vector field. Figure 4B shows the sink areas in a vector field generated from the network related to one of the time intervals between the nodes in Fig. 3A.

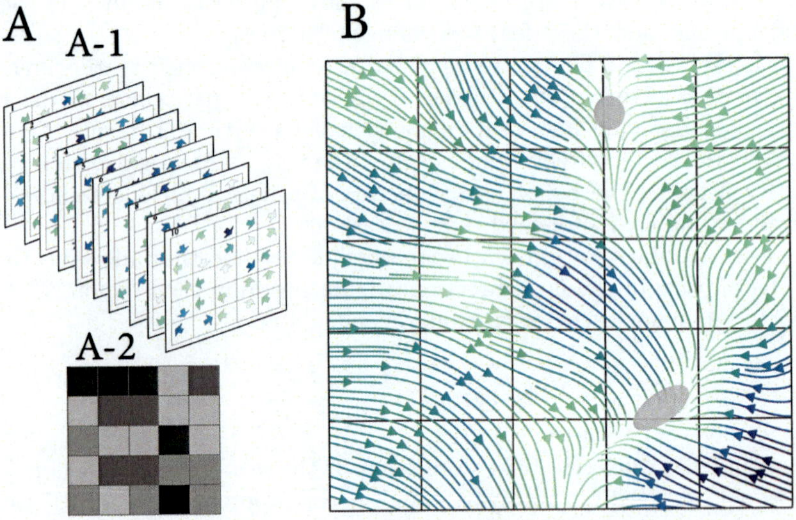

Fig. 4. Clustering and analysis of critical points in vector fields. (A) Using cosine similarity to cluster areas based on the dynamics of their vectors overtime. **(A-1)** ten vector fields derived from mobility network for ten time intervals. **(A-2)** Clustering sub-areas according to vector direction over the ten vector fields; colours indicate clusters. **(B)** Vector-field visualisation, with sinks shown in grey.

4 Case Study: Cattle Trade

In order to look at the proposed methodology in a real scenario, we use the cattle mobility dataset from the state of Minas Gerais in Brazil [9]. In this dataset, every cattle trade movement is recorded including information about the origin and destination of the movement, the purpose of the trade, the date of the transaction, number of animals moved, and premises identification. This dataset covers a period of four years from 2013 to 2016.

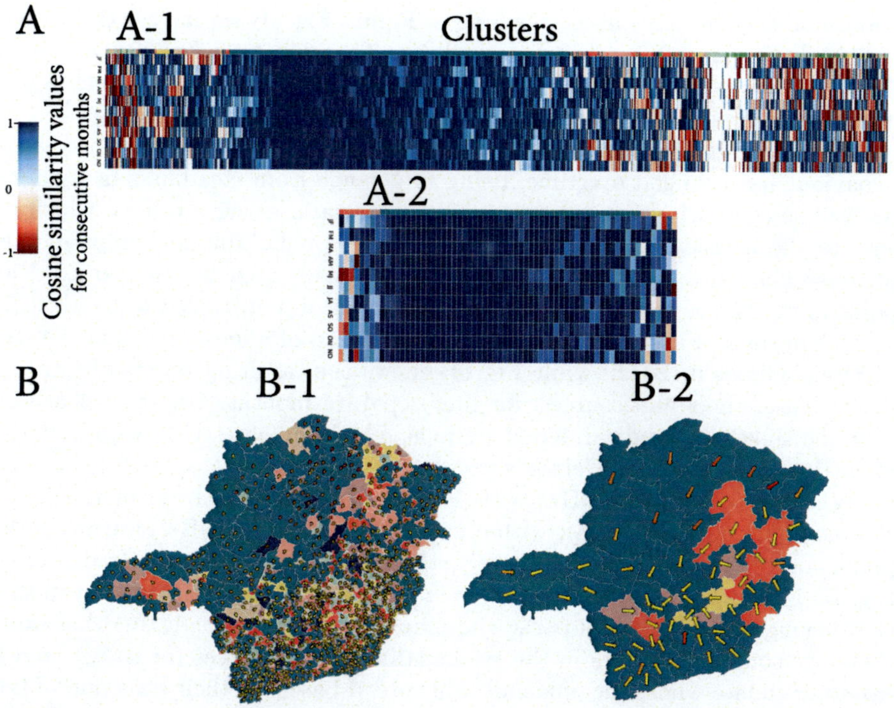

Fig. 5. Cosine similarity between vectors of (A-1) cities and (A-2) micro-regions for eleven consecutive months of 2013. As an example, two flow maps of two consecutive months Jan-Feb are used to calculate the similarity of the angle between two vectors associated with each city. The result of the cosine similarity calculation for each city is a set of values ranging from -1 to 1. Using k-means clustering, cities are grouped according to their similarity values (a feature vector). Each group is illustrated with the same colour in (B). **(B)** Clusters are shown in different colours. The largest cluster of cities **(B-1)** and micro-regions **(B-2)** contains the group that does not exhibit a significant change in direction of the vectors.

We used the proposed methodology (described in Sect. 3) to cattle trade. Due to space restrictions in this work, we use only one year here (2013) to build networks, generate vector fields, and model trade patterns; results for other years have been generated and will be made available in future works. Vector fields are generated using the monthly trades for two different spatial subdivisions, cities, and micro-regions.

A cosine similarity analysis was conducted for cities as well as micro-regions during the year 2013 in order to demonstrate the month-to-month dynamics. The cosine similarity values for eleven consecutive months (Jan. to Dec.) are shown in Fig. 5A for all cities (Fig. 5A-1) and micro-regions (Fig. 5A-2). The values of the similarity are used to cluster the cities or regions as having similar dynamics within the year; the different colours in Fig. 5B indicate different cluster based on cosine similarity. There is a dominant group in both cases, cities (Fig. 5B-1) and micro-regions (Fig. 5B-2). Looking at Figure Fig. 5B we can see that several regions have values close to 1 (dark blue). That happens because the vector direction for the cities (or micro-regions) remains nearly unchanged over a year, indicating a predictable direction for cattle trading from those regions.

The cosine similarity measure shows how a city's (or micro-region's) vector behaviour has changed overtime; being in a range from steadiness or dynamic. As well as comparing the behaviour of one spatial area with others, which can lead to the division of larger spatial areas into predictable and unpredictable regions. The next step is to determine whether an area belongs to a critical point in the vector field. In the vector fields, critical points need to be identified for this purpose. Figure 6A illustrates a vector map of micro-regions in the state of Minas Gerais in Brazil while Fig. 6B shows the resulting points of interests, in this case, sinks and sources, for four separate months. One can clearly see that the sinks change for different periods, which indicates that high dynamics of cattle trades in Minas Gerais is present.

When comparing Fig. 5B-2 with Fig. 6B, most of the areas of the largest cluster of micro-regions (those with bluish cyan color in Fig. 5B-2) are not critical points in vector field. Comparatively, the areas with high sink (red areas in Fig. 6B-1) and source (dark blue areas in Fig. 6B-2) density correspond to those that belong to the cluster of parts whose behaviour is not steady. In other words, there are some risky parts in the system that act as sources (or sinks), and we cannot estimate where the epidemic will spread based on their direction of flow. Even though we cannot predict the exact direction in which diseases spread, we can restrict trade by knowing where the source points of flow are for each period

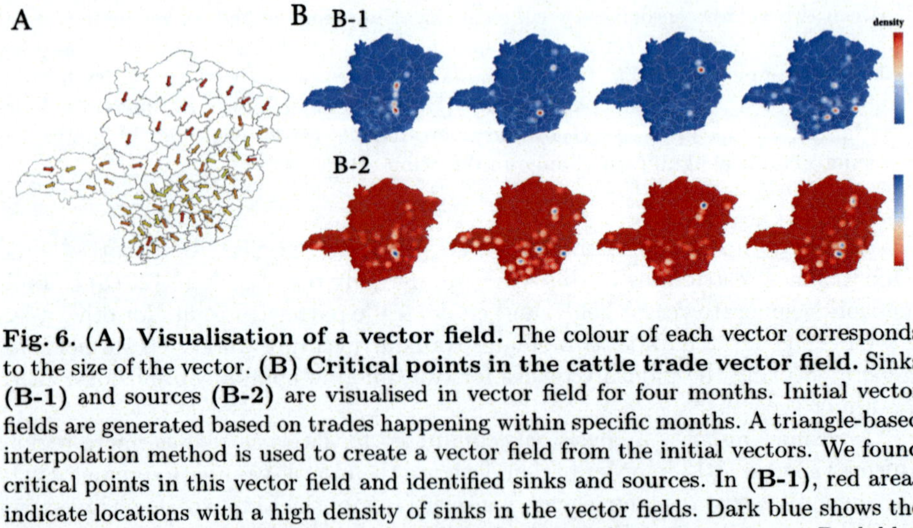

Fig. 6. (A) Visualisation of a vector field. The colour of each vector corresponds to the size of the vector. **(B) Critical points in the cattle trade vector field.** Sinks **(B-1)** and sources **(B-2)** are visualised in vector field for four months. Initial vector fields are generated based on trades happening within specific months. A triangle-based interpolation method is used to create a vector field from the initial vectors. We found critical points in this vector field and identified sinks and sources. In **(B-1)**, red areas indicate locations with a high density of sinks in the vector fields. Dark blue shows the areas with zero density of sink points. In **(B-2)**, the emphasis is on sources. Dark blue areas indicate high-density of sources, and dark red areas are the ones empty of source points.

of time. The next step in future work is to analyse the pattern of changes over time in sinks and sources areas.

5 Conclusion and Future Work

We developed an approach that believe can be applied to any type of mobility dataset which contains origins-destinations trajectories. We showed that the move from networks to flows may be beneficial to the understanding of risk areas and points of interest, such as sinks and sources. Our method provides an alternative/complementary tool to network methods for analysing the dynamic patterns of mobility, and it captures the role of intermediate locations reached between origin and destination of moving entities. A strong aspect of this approach is the flexibility in focusing on outgoing or incoming mobility to each part, as well as the availability of many techniques for combining trades originated/destined from/to each part to produce a final vector for that part based on the investigation goal and application. As a result, it becomes more relevant to a particular type of mobility dataset and analysis goals; for instance, one could consider the number of cattle heads being transported as a factor in the value of the vectors.

We have used an example based on cattle movement in Minas Gerais, Brazil to show how the methodology works in real world scenarios. We intend to do the same analysis of flows for incoming edges in the same context but also look at how effective our approach can be in areas such as human mobility [4]. We expect that vector fields can bring more clarity to temporal patterns in human behaviour, particularly in urban environments.

Despite our application to a real dataset, this work would benefit from a confirmation with local authorities in Minas Gerais that the identified critical points indeed correspond to areas in which issues or risks have been noticed in the past. We intend to continue to work with people in Minas Gerais to reach that level and answer questions related to risk such as: are sources indeed found to be locations in which epidemics start? Are sinks found by our method, locations with a higher change of being affected by a disease spread starting at random locations?

Acknowledgments. The authors would like to thank researchers from the Federal University of Lavras, Minas Gerais, Brazil for the dataset of cattle trade. In particular, Christiane Rocha and Denis Cardoso.

References

1. The biggest producers of beef in the world (2022). https://agroninja.com/countries-where-beef-production-is-defining/. Accessed 21 Nov 2022
2. The laundering of cattle (in Portuguese) (2022). https://piaui.folha.uol.com.br/materia/lavagem-da-boiada/. Accessed: 21 Nov 2022
3. Bajardi, P., Barrat, A., Natale, F., Savini, L., Colizza, V.: Dynamical patterns of cattle trade movements. PloS One **6**(5), e19869 (2011)

4. Barbosa, H., et al.: Human mobility: models and applications. Phys. Rep. **734**, 1–74 (2018)
5. Bates, T.W., Thurmond, M.C., Carpenter, T.E.: Description of an epidemic simulation model for use in evaluating strategies to control an outbreak of foot-and-mouth disease. Am. J. Vet. Res. **64**(2), 195–204 (2003)
6. Beggs, J.M., Timme, N.: Being critical of criticality in the brain. Front. Physiol. **3**, 163 (2012)
7. Bigras-Poulin, M., Thompson, R., Chriél, M., Mortensen, S., Greiner, M.: Network analysis of Danish cattle industry trade patterns as an evaluation of risk potential for disease spread. Prev. Veter. Med. **76**(1–2), 11–39 (2006)
8. Boyandin, I., Bertini, E., Lalanne, D.: Using flow maps to explore migrations over time. In: Geospatial Visual Analytics Workshop in Conjunction with the 13th AGILE International Conference on Geographic Information Science, vol. 2 (2010)
9. Cardoso, D.L.: Cattle Trade Movements: Formation of Patterns, Contact Chain and Epidemiological Risk Analysis (Minas Gerais, Brazil from 2013 to 2016). Ph.D. thesis (2019)
10. Corcoran, J., Li, T., Rohde, D., Charles-Edwards, E., Mateo-Babiano, D.: Spatiotemporal patterns of a public bicycle sharing program: the effect of weather and calendar events. J. Transp. Geogr. **41**, 292–305 (2014)
11. Darbon, A.: Network-based assessment of the vulnerability of Italian regions to bovine brucellosis. Prev. Veter. Med. **158**, 25–34 (2018)
12. Farokhnejad, S., Cardoso, D., Rocha, C., da Mata, A.S., Menezes, R.: A data-driven approach to cattle epidemic modelling under uncertainty. In: CompleNet. Springer, Heidelberg (2022). https://doi.org/10.1007/978-3-031-17658-6_5
13. Ferguson, N.M., Donnelly, C.A., Anderson, R.M.: The foot-and-mouth epidemic in great britain: pattern of spread and impact of interventions. Science **292**(5519), 1155–1160 (2001)
14. Gibbens, J., Wilesmith, J.: Temporal and geographical distribution of cases of foot-and-mouth disease during the early weeks of the 2001 epidemic in great britain. Veter. Rec. **151**(14), 407–412 (2002)
15. Helman, J.L., Hesselink, L.: Visualizing vector field topology in fluid flows. IEEE Comput. Graph. Appl. **11**(3), 36–46 (1991)
16. Hoscheit, P., et al.: Dynamical network models for cattle trade: towards economy-based epidemic risk assessment. J. Complex Netw. **5**(4), 604–624 (2017)
17. Knific, T., Ocepek, M., Kirbiš, A., Lentz, H.H.: Implications of cattle trade for the spread and control of infectious diseases in Slovenia. Front. Veter. Sci. **6**, 454 (2020)
18. Laidlaw, D.H., et al.: Comparing 2D vector field visualization methods: a user study. IEEE Trans. Vis. Comput. Graph. **11**(1), 59–70 (2005)
19. Motta, P.: Implications of the cattle trade network in Cameroon for regional disease prevention and control. Sci. Rep. **7**(1), 1–13 (2017)
20. Natale, F., et al.: Network analysis of Italian cattle trade patterns and evaluation of risks for potential disease spread. Prev. Veter. Med. **92**(4), 341–350 (2009)
21. Payen, A., Tabourier, L., Latapy, M.: Spreading dynamics in a cattle trade network: size, speed, typical profile and consequences on epidemic control strategies. PLoS One **14**(6), e0217972 (2019)
22. Rautureau, S., Dufour, B., Durand, B.: Vulnerability of animal trade networks to the spread of infectious diseases: a methodological approach applied to evaluation and emergency control strategies in cattle, france, 2005. Transbound. Emerg. Dis. **58**(2), 110–120 (2011)

23. Schirdewahn, F., Lentz, H.H., Colizza, V., Koher, A., Hövel, P., Vidondo, B.: Early warning of infectious disease outbreaks on cattle-transport networks. Plos One **16**(1), e0244999 (2021)
24. Sharpley, R., Craven, B.: The 2001 foot and mouth crisis-rural economy and tourism policy implications: a comment. Curr. Issues Tour. **4**(6), 527–537 (2001)
25. Shirley, M., Rushton, S.: Where diseases and networks collide: lessons to be learnt from a study of the 2001 foot-and-mouth disease epidemic. Epidemiol. Infect. **133**(6), 1023–1032 (2005)
26. Smolik, M., Skala, V.: Vector field interpolation with radial basis functions. In: Proceedings of SIGRAD 2016, Visby, Sweden, 23–24 May 2016, pp. 15–21. No. 127, Linköping University Electronic Press (2016)
27. Thornton, P.K.: Livestock production: recent trends, future prospects. Phil. Trans. Roy. Soc. B Biol. Sci. **365**(1554), 2853–2867 (2010)
28. Thornton, P.K., et al.: Mapping climate vulnerability and poverty in Africa (2006)
29. Ungar, A.A.: Barycentric calculus in euclidean and hyperbolic geometry: a comparative introduction. World Scientific (2010)
30. Ware, C., Plumlee, M.D.: Designing a better weather display. Inf. Vis. **12**(3–4), 221–239 (2013)
31. Watson, D.F., Philip, G.: A refinement of inverse distance weighted interpolation. Geo-processing **2**(4), 315–327 (1985)
32. Watson, D., Philip, G.: Triangle based interpolation. J. Int. Assoc. Math. Geol. **16**(8), 779–795 (1984)
33. Wiratsudakul, A., Sekiguchi, S.: The implementation of cattle market closure strategies to mitigate the foot-and-mouth disease epidemics: a contact modeling approach. Res. Veter. Sci. **121**, 76–84 (2018)
34. Woolhouse, M., Shaw, D., Matthews, L., Liu, W.C., Mellor, D., Thomas, M.: Epidemiological implications of the contact network structure for cattle farms and the 20–80 rule. Biol. Lett. **1**(3), 350–352 (2005)
35. Zhang, W., Yang, X., Song, Q.: Construction of traceability system for maintenance of quality and safety of beef. Int. J. Smart Sens. Intell. Syst. **8**(1), 782 (2015)

Author Index

A. S. Teixeira et al. (Eds.): CompleNet 2023, SPCOM, p. 169, 2023.
https://doi.org/10.1007/978-3-031-28276-8